Ta Da!

by

Mary F. Kohnke

New Dimensions Projects

Ponte Vedra Beach, Florida

Ta Da!

This book is a work of fiction. Any resemblance to any person, living or dead, is completely coincidental.

For information contact New Dimensions Projects,
PO Box 1213, Ponte Vedra Beach FL 32004.

Visit ***www.marykohnke.com***
for additional information and to order.

Published 2012
Printed in the United States of America
09 08 07 10 9 8 7 6 5 4 3 2 1

Cataloging-in-Publishing Data
Ta Da! / Mary F. Kohnke - 1st. ed.
p. cm.

ISBN - 13: 978-0-9774536-5-8
ISBN- 10: 0-9774536-5-8

1. German Shepherd- Florida- Fiction
2. Cats and Birds-Fiction
3. Telepathy Animal to human-Fiction
4. Intracoastal and Beaches-Florida-Fiction
5. Animal Rescue-Florida-Fiction

PS3611.0369
Library of Congress Control Number 2012908797

Cover Art and Illustrations

Laurie Allen Klein's cover and illustrations have again brought Duchess, Duke, Tiger, Kitty, and their new friends Dane, Bejee, Hagan, Pip and Bo to life on the pages of TaDa.

Her talent, creativity, and insightfulness have made them come alive on the pages just as they were in my mind. She brings the action and emotions in the stories to life in a very believable way.

When I read and see them on the page I laugh and sigh and say, thank you Laurie — you've done it again.

She can be reached at *LaurieKleinArts.com.*

Book and Cover Design

Kip Williams, Print-Ink Press — *mrkipw@gmail.com*

Also by Mary F. Kohnke

Let's Talk

Dr. Mary Paul in Florida Mysteries:

She Said "What?!"
She Said "A Landfill Where?!"
Murder? In Ponte Vedra?!
The Commissioner

Acknowledgements

The Ponte Vedra Beach writers group saw everything first in its rough form.

Capt. Kerr can't help editing as he reads as well as adding comment, but Bill West drove me crazy with his constant demand for more and more action. It's his fault that Duke had a run in with a rattlesnake but also for making them do more than just talk to me.

A writers group is a real blessing to an author. I heartily suggest it to all, if possible.

Barbara Ames did the final editing. I think she did a special job because of her fondness for Tiger. She genuinely loved the animals and all that happened to them. Makes an author smile.

Linda Oakley and her Golden Doodle, Georgia are very much with us and Duke is still smitten with her. It is so easy to write with these real life models of excellence. Another one joined us Betty Jean Osteen and her Jack Russell Terrier, Hagen. The story is true and the personality of Hagen quite true to form, I believe.

Finally Duke, Duchess, Tiger and Kitty are really glad that they have so many adventures and new friends. They, along with me want you to know we alone are responsible for all the content and any missteps. When I finished the book they wanted to know, "now what?"

I said, "Now Nothing, that's it."

I heard of chorus of, "No way. We have so much more to do."

So, my friends, be prepared for Dog book three in the ... future .

Preface

Welcome to this second book, "Ta Da," as you learn more about Duke and Duchess, the German Shepherds; Kitty, a Calico cat, and Tiger, a Desert Lynx cat.

In Book 1, *Let's Talk,* the narrator, while looking for a German Shepherd dog to adopt, finds not only one but two who can communicate (telepathically) with her.

Later, while adopting a kitten for Duke, she finds a Desert Lynx cat, Tiger, who convinces Duchess and the narrator to adopt him. And so the "family" grows to four remarkable creatures. *Let's Talk* took us through the early growing pains of this new family.

They live in a house set on a large piece of land covered with trees, by the Intercoastal Waterway, with a large dock and boats. The living quarters of the house are on the second floor, with recreation areas and a carport on the ground floor.

The animals have the run of the place, without invitation, except for the bedrooms and office-cocktail area, which has white carpeting, where clean paws are a prerequisite for an invitation.

Book 2, *Ta Da,* introduces us to the new friends they have made and the many exciting adventures they enjoyed as they rescued birds, deer, and babies. They caught thieves and even took their first plane ride.

Duchess had been chosen first, or as she says, she did the choosing. She's a bit vain, somewhat arrogant, insightful, and grumpily kind.

Duke is big and gentle unless he senses a threat to the

others. Then, he turns into a ferocious guardian.

Tiger is funny, clever, and bright. He is always up for an adventure as long as it doesn't involve a trip to the vet or getting into the water.

Kitty has developed an interior life. One might even call it spiritual. She can also talk with almost all God's creatures, especially birds.

The joy of a telepathic family, for the narrator, is double-edged. It allowed her to set clear rules, such as where one does their business, where one does not scratch or chew or sit, and one does not leave the yard nor make unnecessary noise. These clearly stated directions are offset with endless questions starting with the word *"why?"* or with words like *"can I?"* or *"can we?"*

Each day unfolds as a unique singularity in lives that are experienced fully in the *now.*

As you open this new door, *Ta Da,* into the lives of Duchess, Duke, Kitty, and Tiger, you find them talking telepathically with their owner.

So, please… after you!

Table of Contents

1
Help, Help!

"Can you help me save these little fish?" Kitty called out.

We were at the beach very early in the morning during the falling tide. A small pool of water had formed and sand blocked it from running out to the ocean.

"See," Kitty said, anxiously pointing a paw, her little tail sweeping back and forth with nervous anticipation, *"if they can't get out to the ocean they will die when the water sinks into the sand."*

Tiger and I looked it over. He said, *"If you take your towel and sink it into the water we can chase them into it and you can carry them into the ocean."*

I looked, *"OK, that'll work."* I sank the towel to the sand bottom. It was only three inches at the deepest part. Kitty sort of danced the fish toward the towel, while Tiger high stepped it. Like he wasn't in the water if he stepped high.

"It's working," Kitty said. *"See, they're all above the towel. Now pick it up and carry it out into the water and let them go."*

As I waded back to shore, I saw a man walking fast near the dunes, heading in the direction of the parking lot, carrying a small child. He was thin with longish hair and baggy work pants. He saw me watching, waved and yelled in a high pitched voice, "Nap time for this one."

"I wonder where he came from," Tiger said. *"I didn't see anyone but us, and Duke and Duchess out here."*

"We were kinda busy, weren't really looking. See Duke and Duchess are way up there. They sure can move. Bet they're having a race," Kitty said.

They were running and splashing in the water near shore. They loved the beach. I did too, despite the baths afterward to rid them of sand and salt. Now I had to do Kitty and Tiger too. That should be fun, I thought to my self.

"Help, Help!"

I looked around. A woman was rushing from one of the houses above the dunes. Again I heard her yell. "Help, Help! My baby." Her voice was tinged with alarm, fear and anxiety. She was red faced as she came down a path thru the dunes. "He took my baby," she yelled.

I looked in the direction the man had gone with the child. I could just see him starting up the walkway to the parking lot.

"Get Duke and Duchess quick," Tiger yelled.

We all turned and I yelled, "Duke, Duchess come. Hurry."

The woman was running toward us. She hadn't yet seen the man going up the walkway. She heard me yell for the dogs slowed and looked back at them. Then she turned and saw the man.

"We have to stop him. He has my baby." She started going after him.

Duke and Duchess came to a splashing stop by us. *"Hey watch it, that's wet,"* Tiger said jumping back.

"What's going on?" Duchess asked, looking after the woman who was running away.

"A man stole her baby. You've got to get it," Kitty said.

Duke looked at me, *"What baby?"*

"You two have to go after the man now. He's headed toward the parking lot with a small child. Hurry! Be careful," I said.

They were off like a shot, racing past the woman. I wanted to go after them, but I wasn't leaving the cats on the beach alone. *"Come on, I'll carry you. Tiger you get on my shoulders"* I shouted as I picked him up, *"but watch those claws,"* I picked up Kitty and started after them.

"They'll catch him," pronounced Tiger. He had draped himself around my neck and his hind legs were on one side and his front legs on the other. He had one paw and claws in my shirt. I jogged up the beach.

"If you will hold Kitty under one arm and grab my rear legs with the other hand we can balance better and go faster," Tiger called into my ear.

I tried. It worked.

The woman in front of us slowed, and turned. I yelled, "Did you call the cops?"

"Oh my God! No!"

"Do it now." I yelled..

Ripping her cell phone from her slacks pocket she dialed 911 and explained what was happening while jogging along side us. "Are those your dogs after that man?" she asked.

"Yes, tell the cops they're German Shepherd police dogs."

"Wow!" Tiger said. *"That's impressive and kinda scary."*

"They won't get hurt will they?" Kitty asked.

"I hope not" I replied.

"I'll pray hard. You too, Tiger."

* * *

Racing across the sand, Duchess called, *"How do you want to do this, Duke?"*

"I'll get the man, you go for the child."

"Look! He's going up the walkway. He can't go too fast carrying that child."

"Yeah, I see him, Duchess. I don't think he's seen us yet."

"Let's stay on the sand. It's lower then the walkway. They both go to the parking lot, but we won't be as noticeable down there," Duchess said. *"I hope he hasn't got a gun."*

"Me too," echoed Duke.

The man jumped off the walkway and went across the parking lot toward a dark blue pick- up truck. He wrenched open the door and grabbed a large beach towel off the seat. The child was twisting around, kicking and crying. The man wrapped the towel over the child, holding it secure and muffling the screams.

"There he is. He's covering the child," Duke said, still going at break neck speed. *"I'll get him, you get the child."*

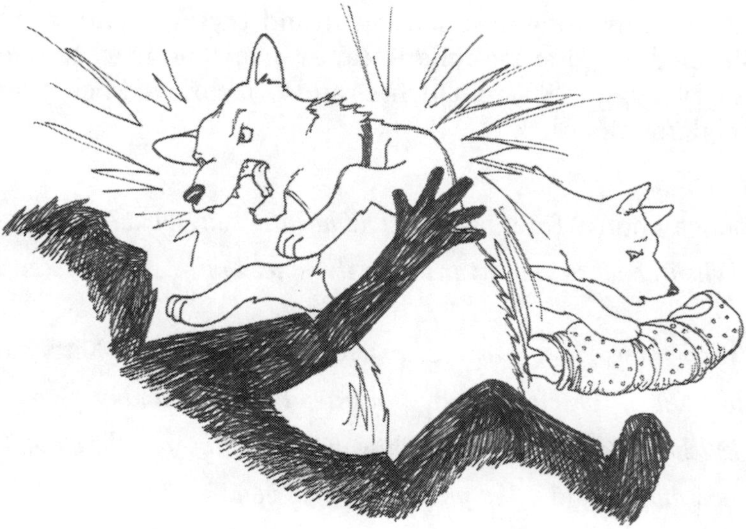

Duke hit the man with a full body block knocking him back into the door and the child out of his arms onto the ground. He was growling, snarling , and biting. He grabbed the man's arm and shook him.

Duchess grabbed the towel, pulling towel and child across the lot away from the car toward the trees and bushes by the beach.

The man kicked Duke away. He grabbed a big flashlight off the floor, swinging it at Duke, driving him back. He looked up and saw the child being dragged by another big dog. He started toward them still swinging at Duke when sirens could be heard in the distance. Swearing, he swung once more at Duke turned and ran for the truck.

"Let him go, Duke", Duchess yelled. *"Get out of the way of the truck."*

Duke jumped aside as the truck raced out of the lot. He walked over to Duchess and sat. panting. *"Wow, that was something. How's the child? Is it still crying?"*

"Yeah, still scared." She pushed the towel away from the childs face and gave it a lick.

Then Duke nosed in and licked it. The child hiccupped a couple of times, sniffled and reached out a hand and touched the dog's fur. "Doggy."

"Licking works every time with babies," Duchess said.

"Yeah, worked with Kitty when she was a little," replied Duke.

They sat close on each side of the child, pushing with their noses and giving an occasional lick. That's what we saw when we got to the lot.

The mother rushed over to the dogs, knelt next to them and took the child into her arms, hugging, kissing it and crying joyfully. When the mother stopped hugging the child it leaned away and reached over to the dogs. "Doggy, doggy," it said. Duchess and Duke moved nearer and both gave the child another lick

* * *

On the way home Tiger asked, *"Did he hurt you, Duke?"*

"Na. When he kicked me I turned so it glanced off me. He missed with the flashlight."

"You're so brave, Duke. I was really worried. We all were," Kitty said.

"Me too, Duke," Duchess said, *"I wanted to go back and help you."*

"No, Duchess. You did the right thing and we both saved the child."

"Yes, you did," I said. *"I'm proud of both of you."*

"Did the woman tell you who that man was and why he tried to kidnap her

child?"

"Yes. She said he was a yard man she'd fired a couple of weeks ago for stealing. I guess he thought he could get money from her for the child."

"Good thing we were at the beach today," Tiger said. *"We saved the little fish and a child. Uh-huh, yes we did,"* he said, nodding his head.

2
Parade

Kitty has been practicing riding on Duke's back, but when he turned or moved to fast she had to dig her claws in to hang on. It was Duke's idea for her to ride so she could go with them when they ran in the woods. But, you could tell it hurt him.

Duchess told me, *"You need to get him something to wear for Kitty to hang on to."*

* * *

At Duke's fitting, I had the man put extra padding on the top of the vest. He didn't ask why I guess he really didn't care.

Duchess had on her fancy red collar with sparkly stones. She watched the whole proceedings, than said, *"Do you suppose I could get one, too?"*

Uh-Oh, I thought. *"You want to take Kitty for a ride, too?"*

"That would be nice, Duchess. Kitty would like that," Duke said.

"Yes," said Duchess. *"I want mine to look like my dress-up collar."* She turned around, showing off the collar.

"Please?" Duke asked me.

I told the man what we wanted. Duke's in black leather, with a bit of red trim; Duchess' in red leather, with a bit of black trim, plus some sparkly stones.

* * *

I was on the phone with Linda. Her dog is a 'Golden-Doodle'. Or as her dog Georgia would say *"I'm a designer dog - very pretty and very expensive."* Or as Duchess said, when she first met her *" A ...mixed breed."*

"So Georgia's going to be in the parade of dogs."

"Oh yes," replied Linda. "She'll be a hit, in a lovely pink tutu".

"A pink tutu!" I repeated with a bit of astonishment and a smile in my voice.

"What? You don't like it?" She shot back from the phone

"No, No Linda, it sounds wonderful. She'll be beautiful, a showstopper. When is this? Can I come and watch?"

Duke, Duchess and Tiger were dozing by the door of my office. They suddenly sat up, ears perked and alert, avidly listening. The sound of Georgia's name really caught their attention.

When I put the phone down I heard, *"When is this? Can I go and see Georgia?"* Duke asked.

"What's a tutu?" Tiger asked.

"It's a dancer's costume. It has a top and a short stiff skirt that stands out from the body," I replied

"She's going to wear that? Oh boy," said Duchess with much eye rolling. *"No self respecting dog would ever do that."*

"I bet it will be pretty," said Kitty, who had joined them at the door.

"Yes Kitty. Pink will be very pretty with Georgia's blond fur. Oh, I so-o-o want to go and see her," Duke said.*"Can I, Can I?"*

There they sat. Duke with a pleading look, Duchess looking disgusted at him and at the idea of a tutu, Kitty smiling and Tiger just being.

"Well, I know we can go. It's a fund raiser."

"What's that 'fund stuff'?" Tiger asked.

"It's when you put on a show and people pay to come see it. This money will go to a shelter that rescues dogs to help pay for their food. Duke and Duchess can come with me but I'm not so sure about you and Kitty. There'll be lots of dogs there, so you can't be on the ground walking about."

"Kitty can ride on me. On my new vest," Duke said nodding.

"You think that will be safe?" I asked.

"Ohhh Yes. No one will dare come near me. If they get too close I'll tell them No. I can also smile at them." He said showing a lot of white teeth.

"Don't worry. Duke and I have been with other dogs. They do whatever we tell them," said Duchess, with a turn and lift of her head.

"Good, it's all settled." Then Duke looked down at Tiger. *"Well, almost settled."*

"No, it's OK. I don't have to go." He turned and left the room.

Duke looked at Duchess and waited.

"What are you looking at me for?" she asked. She looked at Kitty and me. Then back at Duke. *"He's big enough to walk. He has a harness and leash."*

"I don't think so, Duchess," Duke quietly said. *"He wouldn't be safe on the ground."*

Duchess looked at me. *"You could carry him."*

"No," interrupted Duke. *"Dogs aren't afraid of people carrying cats. Won't work,"* he said shaking his head.

Duchess stared at Duke, looked at Kitty then me. Finally with an air of disgust, she turned and walked out.

"Is she mad at us, Duke?"

"No Kitty. I think she feels trapped. She'll work it out."

* * *

The next day Duke and Duchess were walking about wearing their vests. *"Getting used to them,"* they said. I saw Kitty riding on Duke. They were practicing runs, turns and fast stops. They stopped and were looking off to the side of the house. I walked out on the porch to look too and there was Duchess walking slowly with Tiger riding on her vest.

"Am I too heavy. Duchess?"

"You're no feather."

"Would it help if I lay down? You know sort of spread out."

"OK, try that.... Humm. Move forward a bit.... OK, try crouching." Duchess moved about and did a few turns.

"I think when I'm walking faster and turning that crouching is better. But when we're standing still, sitting will be OK. Move up closer to my front legs when you sit.... Yeah, that's better."

"Can I practice jumping on and off a few times?"

"Good idea. Cats are good jumpers, but don't miss the vest and get me with your claws."

"Oh no, Duchess. I'd never do that," Tiger replied.

* * *

The next Saturday we arrived at Marsh Creek, for the parade fundraiser. Kitty and Tiger mounted up and off we went to find the place where Linda told us to sit. The looks we got were priceless. Linda had Georgia on a leash and waved us over.

Duke walked up to Georgia, *"Oh, you are so-o-o pretty, Georgia."*

"Thank you," she said, as she preened around in a circle. Showing off her outfit.

Duchess said, *"Yes, Georgia, you look....very nice."*

Georgia looked at her, nodded hesitantly. *"You look lovely too, Duchess."*

"Yes, I do." She turned in a circle with Tiger sitting high on her front shoulders.

Linda stared. "My word! They're color coordinated." She looked at Duke. "Duke and Kitty too. How... lovely."

"She doesn't look too happy," said Tiger.

"That's cause we're showstoppers," Duchess said, with a toss of her head.

"Oh no, Duchess. You are regal, no question. But Georgia is the showstopper," Duke said rubbing his head on Georgia.

A huge Great Dane walked up, dressed like a bunny rabbit, with long white fuzzy ears and a tail. He stopped in front of Duke. *"What's with the cats?"*

"They're my friends, Kitty and Tiger," Duke answered.

"Good, drop'em and we'll chase them. Be fun." He looked around. *"Lots more fun then this dumb parade."*

Duchess rose to her full height. *"No. That's not going to happen. Ever. Got it?"*

The Dane looked her over and then Duke. *"OK, no chase."* He hung his head, *"I just feel so foolish with these fuzzy ears and the tail. They take away my dignity"*

"Oh no, Mr. Dane. Nothing could ever take away your dignity. You are the tallest and most stately dog in the whole world," Kitty said, looking way up at the Dane. *"Mr. Dane, you do your stately walk, and Georgia, you do your high stepping prance. No one will notice what you're wearing. They'll only admire you."*

"Really? Thank you." The Dane looked at Georgia, *"I'll walk behind you in the parade and we'll strut our stuff. We'll show them. Kitty, if anyone tries to bother you or Tiger, I'll help Duke and Duchess teach'em some manners."*

There were other dogs nearby, watching and listening. They nodded respectfully to Duke and Duchess and walked past.

It was a wonderful parade. The Dane and Georgia both won ribbons. Georgia's was pink and the Dane's was purple. *"Which was best?"* asked Tiger

"They were all best," Kitty answered.

3
Stair-man

I was on the dock with Tom, explaining where I wanted the stairs. "They must reach from the sand to the top of the bulkhead."

"OK, I'll take my measurements and get the lumber," Tom replied.

"Do I need to put the dogs up while you work?"

"Hey, what's with this, 'putting of dogs up?" Tiger asked.

"No, I like animals."

"Yeah!" said Duchess. *"So we're staying,"* she added, nodding her head. She learned this head nodding from Tiger.

The stairs were for the animals. In case they fell or got pushed into the water. The bulkhead is almost six feet high. It happened once. I luckily used the neighbor's low floating dock to boost Duke up. But the neighbors moved and the dock was removed.

* * *

I was working at my desk when Tiger ran to the door saying, *"Hey, the stair-man is here and we're off to help."* He spun and was gone.

I looked out the window. Tom set up his boards, sawhorses and tools. Duke, Duchess, Tiger and Kitty were all lined up watching.

"Think he'll let us help?" Tiger asked.

"Well, we can't very well ask him, can we?" replied Duchess looking down at Tiger.

"Oh, you know what I mean," he paused.... *"Duchess. Don't be mean."*

"I think we can find ways to help, if we just wait and watch, Tiger," Kitty said.

"What do you mean?"

"See," Duke said. *"Sometimes he lays a thing down and then has to hunt for it. If we watch and listen we can figure out what he wants and hand it to him."*

"Gotta be careful he doesn't begin to think something is funny with us," Duchess said. *"We already talked about what might happen if people found out*

we can talk with her and each other telepathically. They might steal us and put us in a zoo."

"Yeah, so we will start out slow and smart. Wow! What a game." Tiger's eyes gleamed.

* * *

After all the wood was cut. I walked down and asked how it was coming.

"I cut the two long boards, they are notched for the steps to sit on, there called stringers. The steps are cut and the big header board. I'll nail that to the bulkhead to fasten the stringers to," Tom explained. "The tides out I'll dig the holes for the two short 4x4 pilings. You can hold them in place while I pour the quick-dry concrete around them."

I went down the ladder after the hole was dug I held the piling upright in the hole.

"It's not straight," Duke said. He lay on the crosswalk above my head eyeing the piling.

"It needs to go more toward me," said Duchess, who was lying on the adjacent walkway.

"A bit more toward me, too." Tiger said, who was on the walkway across from Duke.

I moved it a bit toward Duchess. Tom eyed it, "Move it a bit toward that cat up there."

"See, I told you," piped up Tiger

Kitty sat on the bulkhead. They had it surrounded on all sides. *"Yeah, that does it,"* she said. I rolled my eyes, said nothing and followed directions. Tom dug the second hole and we went thru the same, *"a bit toward me," "now a bit toward me,"* and finally, *"yeah now you got it,"* from Kitty.

Tom eyed both pilings and nodded. "You have a good eye," he said. "It usually takes me longer to get them so straight."

"Thank you." I smiled.

"Yeah we did a great job," Duke said.

"Been here forever without us," agreed Duchess.

"See, I told you we could find ways to help," Kitty said, as she washed her face.

"Now what?" Tiger asked. *"What comes next? Ask him please, cause we're ready."*

"What comes next, Tom?" I asked.

"It'll be a couple of hours till the tides in. They'll be solidly in place by then. Nodding toward the pilings, he said, "This quick-dry cement works fast. Tomorrow we'll do the stringers and steps. I'll start when the tide goes out again."

"Good. We'll be here, ready to go," Duke said. The others nodded.

"Those are nice dogs, cats, too. They sure are attentive. Watch every move I make." He laughed. "Never had such close supervision."

"They aren't bothering you, are they?" All four turned and stared at me.

"Bothered? I don't think so," said Duchess. *"We're his helpers, you'll see."* They all nodded.

"No. They're good company."

* * *

I heard the dogs bark. Tom had arrived. I walked out and asked, "Do you need my help with the stringers?"

"Yes, I have the header bolted on the bulkhead. If you'll hold each stringer in place, I'll bolt it to the header and the piling. The rest is easy."

When we finished I heard, *"OK, you're done. We can take over now,"* Tiger said. Three heads were nodding.

"It's OK," Kitty, rubbed up against me. *"We won't get in his way."*

I left.

"Let's watch him and if he reaches for something we can hand it to him." This went on for a while.

Once Tom said, "Where's the darn hammer?"

Duke picked it up in his mouth and brought it to him. "Thank you Duke. You're a smart dog. I guess if I just ask for some thing you can get it for me. Like a game."

"You got that," Tiger quipped.

He was drilling long screws in, anchoring the steps. When he looked up reaching for another, Kitty, standing above him, with one in her mouth, dropped it in his hand. Tiger was next to her with two in his mouth. "Well I'll be darned. It's like you can read my mind."

"Oh, Oh," Duchess said. *"Better be careful, he maybe getting suspicious of us"*

"It's OK," Duke replied. "He *just thinks we're really smart dogs and cats,"*.

"See, sometimes he just holds his hand out and either Tiger or Kitty are right there with a screw. But if he wants something heavier he asked Duke or me. I don't think he's even thinking about what we're doing," Duchess replied.

"I hope not," Duke said. *"We'll find out when she comes down to check on the work, I'm sure he'll say something."*

* * *

"Looks like you're about done, Tom. Did you have a lot of help?"

"You can say that again. Whenever I wanted something, one or the other was right there with it. Almost uncanny, but I have friends with hunting dogs that are that smart. But I never knew cats were that good."

"Really! That's nice." I glared at my pets. I was pretty sure no one would believe we talked telepathically, but they had been pushing it the past few days with Tom.

"We didn't do anything," Tiger said always quick to the defense.

"No," Kitty piped up. *"We made it look… almost normal."*

"Almost?"

"Well yes, Duke and I played… like fetch with him." Duchess said.

Tiger ran over to the steps and looked down. The tide was almost in. *"Can we try the stairs when the tides go out again?"*

A real subject changer, my Tiger. *"Yes, we can do that tomorrow."*

4
Learning to Swim

I was looking out the window toward the dock. Duke, my German Shepherd held Kitty in the air by the nape of her neck, As I watched, he gently put her down on the dock. His sister Duchess was next to him as was Tiger.

I quickly walked out and Tiger looked up, *"Wow, that was close. Duke just snatched Kitty from the jaws of disaster."*

Oh boy." Duchess rolled her eyes. *"You're right. He's watching too much television."*

"What happened?" I asked.

"Kitty was sitting on the edge of the dock looking down into the water," Duchess explained." *The tide was going out. It was shallow. We wondered what she was seeing, so Duke and I walked out to see. It was one of those tall white birds wading in the water just below her."*

"We must have scared the bird because it suddenly just burst up into the air and startled Kitty. She lost her balance and almost fell in," Duke added.

"Yeah, and Duke then snatched…"

"We know, Tiger. He grabbed her so she wouldn't fall in the water," Duchess said

I sat on the dock and put Kitty in my lap. *"Are you OK?"* I asked as I stroked her fur.

"Oh, yes. I felt myself starting to fall and then Duke grabbed me." She looked at Duke. *"Thank you.."*

He nodded and rubbed his nose against her fur.

"I'm sorry, I didn't mean to scare anyone. I asked the bird if it was OK if I watched her fish. She nodded her head. She was really fast. It was fun to watch her. We didn't hear Duke and Duchess till they were almost on top of us."

Kitty could talk with birds and had on a number of occasions.

"I was running up right behind Duchess. Maybe I scared the bird," Tiger said in a small voice with his head downcast.

Kitty shook her head. *"Oh no, Tiger, you run very quietly. I think she just suddenly saw big shapes above her and reacted with flight. That's what birds do you know."*

We sat there quietly.

"Kitty needs to learn how to swim," Duchess pronounced.

"In the water?"

"Yes, Tiger. Where else?" Duchess asked in her haughtiest of voices.

"I don't much like water," Tiger admitted in a soft voice.

"You already know how to swim. We live by the water, and Kitty needs to know how to swim," Duchess said looking at me. *"You have to teach her."*

Duke was very still. He looked at me, then at Kitty and nodded his head. *"Yes, Kitty.*

You need to learn how to swim. I may not be near next time."

"OK Duke." She looked up at me. *"Will you teach me?"*

"Yes, Kitty."

"You could wade out into the intracoastal, at low tide by the dock,, drop her in and let her swim to shore," suggested Tiger nodding his head in agreement with himself.

*"No. The water is salty and not too clean. The bottom is dirty, too,"*Duke said, emphatically.

"Drop her in the bathtub," Duchess said. *"Its clean."*

"No. She should be able to swim to the edge and walk out," Tiger nodded.

"Well?". They asked in unison, staring at me.

"OK. I'll make some calls and find a swimming pool that has stairs coming up out of it."

"Can we all go swimming in the pool too?" Duke asked.

"I'll ask if we can."

"I don't think so," Tiger stated. *"I don't …."*

"We know, Tiger. But how do we know you really can swim? You say so, but if your fibbing to us, just cause you don't like water. "Well… " Duchess looked around at us, *"We sure wouldn't want to find out that you can't the hard way."*

Tiger crouched down, lower his head, looking scared and sad all at once. *"That's mean, Duchess,"* he said softy.

Kitty jumped out of my lap. Went to Tiger, purring and rubbing on him. *"Tiger, don't feel scared. No one's going to hurt you or just toss you into the water. Duchess is just teasing you."*

Duchess looked at Tiger, then at Kitty and the rest of us. "*Maybe.*" She walked away.

* * *

A few days later, we arrived at my friend's pool. They were out of town and assured me we could all use the pool. She said her dogs loved to swim in it. I brought towels for all of us. Kitty sat on the edge and watched Duke and Duchess swim up and down past her. *"See Kitty, how we move our legs. No problem."*

"Yeah Kitty, just jump in and start moving your legs like us," Duchess said

"See Kitty, when you get to the steps you put you feet down and walk out like this." Duke swam to the steps and walked up them. Duchess followed him out of the pool and they both shook water on Tiger who was sitting near the steps.

Tiger jumped a foot in the air, yelling, *"Watch out! That's wet."* and scooted back from the pool, shaking himself.

I couldn't help but smile to myself.

I got in the pool and asked Kitty if she wanted me to hold her under her middle while she practiced. I assured her I wouldn't let her drop deep in the water. She nodded.

I picked her up and set her gently into the water and told her to move her legs like she was walking fast. The dogs were laying on the edge of the pool watching every move. Even Tiger crawled nearer the edge on his belly.

I walked with Kitty toward the steps as she moved her legs for all she was worth. I was tempted to let her go, she was doing so well. But, she trusted me not to. When she felt the steps under her feet she scooted up them on her own, turned, shook and sat on the top step.

"Wow. That was something." She looked at Duke *"I was swimming, wasn't I?"*

"Yes Kitty. You did good."

"OK. But I need to do it without being held." She looked at me. *"Can we try again?"*

"Yes. I'll hold you like we just did. You start swimming and then I'll take my hands away."

When I took my hands away, she dropped under the water a little, came up sputtering and went on swimming to the steps and climbed out. Duke was crouched ready to leap into the water, Duchess as well. Even Tiger was on the very edge of the pool.

Kitty shook herself, looked around, sat and proceeded to lick her paw and wash her face. Just like this was old stuff.

"You OK, Kitty?" Duke asked.

"Yes." She continued licking herself.

I threw balls for the dogs on the lawn near the pool. Duke missed a few, as he kept looking back at Kitty.

"Kitty, were you real scared?" Tiger asked.

"No. I knew all of you were right there. But, I didn't like getting the water in my nose when I went under. I forgot to keep my head up and move my feet at the same time."

"Yeah, that happened to me. Water got in my nose, when those boys threw me in. All I wanted was to get out of there. I, like started running in the water, and it worked."

"It wasn't bad once I got moving. Tiger, I would like to try again. Will you swim along side of me?"

"Well I don't know…" He looked in the water and then at Kitty. *"How far?"*

Kitty looked at the pool. *"We could swim across this narrow end to the steps across the pool."* She looked up at Tiger. *"We can swim right near the edge so if we get scared we can grab on."*

Tiger looked at the edge and across the pool to the steps. *"OK. But just once."*

And in they went.

"Look!" Duke exclaimed.

"Well… I guess he can swim," Duchess said. *"Kitty, too."*

5
Watch Out

"Kitty says hurry, real fast!" Duke yelled from the door.*" Bring the boat keys and the scissors. Hurry!"*

As he turned, I yelled, *"Wait a minute, Duke. What's going on?"*

"That bird I rescued is back. She's on the dock. She told Kitty her mate is trapped in netting and will die. So we gotta hurry." He raced off.

Some months ago a cormorant got caught in netting, out beyond our dock. Duke jumped in the water and brought her to the ladder. She fought us when I tried to cut her loose till Kitty told her to be still. That's when we learned that Kitty could talk to birds.

I grabbed the scissors and boat keys. When I got to the dock to lower the boat, they were all sitting in the bow waiting. The cormorant was perched on the dock waiting as well. *"Which way do we go Kitty?"*

"She said that way," Kitty pointed south with her paw. *"She'll lead us. She said it's her mate and they've been together a long time."*

"What's a mate?" Tiger asked.

The others looked just as quizzical. "It's *another name for a mother or father who get together to have babies."*

"How do they...."

"Never mind Tiger, lets go."

All four were sitting in the bow on the seats. The bird was flying in front of us.

She flew past the Palm Valley Bridge. The tide was rising. A mile or so further she started to circle then dropped and landed on the shore. I slowed the boat and could see a bird struggling in the water near the shore.

"You guys stay in the boat." I beached the boat on the shore a bit, grabbed the anchor, got in the water and stuck it in the sand.

"Kitty, tell her to tell him to stop struggling so hard."

"Yeah, and tell him not to bite her," Duke said. *"She almost got us when we saved her."*

"What's he tangled up in?" Duchess asked.

I walked over near the bird. *"He's caught in both netting and a piece of rope from that crab trap, over there."*

Both the female bird and Kitty were talking to the male bird. I waded out a few feet and reached down to pick the bird up and yanked my hand back in a hurry.

"Hey!" Tiger yelled, *"Watch it!"*

"Yeah," Duchess echoed. *"Maybe we ought to leave him here."*

"No. It's just really scared. Maybe if you just start cutting the rope and loosing the net slowly he'll calm down," Kitty said. *"Then you can pick him up."*

The female swam nearer to male, chattering away.

"What's she saying, Kitty?" Tiger asked.

"She's telling him to be still and not bite or struggle."

"Tell her to tell him, we'll leave him here if he doesn't cut it out," Duchess said.

I cut the rope loose from the crab net then slowly pulled the net and bird closer to shore. The female was right next to me still talking to him. *"It's a real mess. Someone threw their net and didn't see the trap and it got all tied together. It looks like when he dove, he didn't see the net. The more he struggled the tighter it got. Tell him I need to get him on shore so I can cut it from around his feet."*

His head and neck were free but his wings and feet were still tangled. I put the scissors under the net down his back and cut. The net fell away and he flapped his wings splashing water all over.

"I'm sure glad that's you and not me. I'm not much for water," Tiger said, dodging back from the edge of the boat

The bird tried to move his feet. But they were still snug in the net. *"I'm going to have to pick him up to get at his feet. Tell him to hold real still."*

The female almost pushed him into my lap and got her head between his head and my hands. I grabbed his feet and cut away. One final snip and he was free. I set him down and he sort of waddled further up on shore with the female right beside him.

He was about two and a half feet long, with thick dark plumage, pale chest, and green eyes. He appeared really bedraggled, as some of his feathers were out of place and ruffled. He had a long curved neck with a bill that was long and the upper part was hooked down. His feet were webbed. When his wings were extended to dry he was a very pretty bird.

"We did it," Duchess said.

"Yeah, we're really good," Tiger echoed.

"Kitty, you saved another bird. I'm so-o-o proud," Duke said, rubbing his nose on Kitty.

"Will he be able to fly?" Duchess asked.

"I think so," I replied.

"He sure could flap those wings," Tiger muttered.

"Kitty, ask *her if he can fly,"* I said.

They chattered away for a while. The male bird was able to walk. He put his wings out and flapped. Then settled down. *"He's really tired. He'll need to rest a while to get his strength back,"* Kitty said.

I put the anchor in the boat. Standing by the boat a while, I noticed they seemed to be talking together. Finally Kitty turned to me. *"He thinks he's fine and well… sort of wants us to go,"* Kitty said.

"Huh?" Duchess said. *"Dismissed, without even a thank you? The nerve. I told you we should have just left him, all tied up."*

"No Duchess, he's just scared. He's never been so close to people before," Kitty said.

Duchess was sitting in the bow on top of the seats. The others were near her. *"You guys get back a bit so I can climb in. I can push the boat off with the boat hook."*

Duchess moved further toward the far side but still in the front. I climbed up and over the bow and walked back to get the boat hook when I heard a loud, *"Watch out! Wow! look at that,"*

Tiger yelled. *"It's a monster, a sea monster."*

I quickly turned and there was a large gator slowly coming out of the marsh grass south of us. *"Hurry, fly away, fly away,"* Tiger yelled

The alligator looked like an ancient predator. Powerful long jaws filled with teeth, slitted eyes, hide like dark armor over a lizard like body. It was making its way slowly toward us.

The female cormorant lifted up in the air but the male could only waddle and flap its wings. *"It's too tired,"* Kitty cried.

I started toward the bow, measuring the distance from the gator with my eyes. *"You're not going back in, are you?"* Kitty asked, in a tiny voice.

"No. You can't do that," Tiger yelled, visibly upset.

Suddenly, Duchess shot off the boat, landing on the shore near the bird. She took it in her mouth and after a few steps leapt back into the boat. She set it on the floor, stood over it, and said, *"Kitty, you tell that bird not to dare move. Not so much as a feather."*

The bird was shivering with fear. Kitty walked up to it and tried to calm him down. *"All it wants to do is escape and hide."*

"Yeah. Where? Back with the gator? Maybe I should've left him there." Duchess grumbled, not moving one bit from the bird. She was afraid he would get back in the water.

I pushed the boat back and started the motor. I looked back at the gator it had stopped at the waters edge and watched us leave. *"Lost its lunch, it did,"* Tiger nodded. Standing bravely in the bow watching the gator disappear in the distance.

I felt the air stir near my head and feathers brush my shoulder as the female landed near me. She eyed Duchess.

"Tell her to keep him still, Kitty."

There was some chatter *"She will, Duchess."*

We motored slowly along. Finally Kitty said, *"They want to sit up in the bow alone."*

"Hey, that's my place," Tiger said.

"Just for a little while, then they will fly away. She'll know when he can fly," Kitty replied. *"You can go a bit faster,"* she said to me.

What a sight. The animals crowded close to me as I steered. We watched the birds standing with their wings fully spread. I think they were drying them. Soon Kitty told me to go a bit slower. Both birds soared into the air, did a turn or two, then swooped down over the boat and were gone.

"They sure are pretty," Duke said. *"I'm glad you saved him, Duchess. That was so brave."*

"Well, I had to or she would have jumped in." She nodded at me. *"If the gator got her who'd drive us home?"*

"Who indeed?" I mumbled as I stroked Duchess.

6
Golf

I was putting my golf clubs in the car when I heard, *"I wish we could play golf sometime."*

"Yeah me, too," Duchess echoed.

I looked up and Duchess and Tiger were sitting by the car watching me. *"Really,"* I said.

"What would you use for clubs?"

"When you hit the balls onto the green like they do on TV, we could slap them into the hole like we do the pool balls," Tiger replied.

"See?" said Duchess. *"I told you he was watching too much TV. But this time, he's right, you know. We could do that."*

"OK, I can see that, but they don't let dogs and cats run around on the golf course. Bye now. I'll be back later." A glance in my rearview mirror showed them staring at me with their most doleful looks. I focused on the road trying not to feel too sad for them. I'd have to think of something.

* * *

I found a course with no houses around it. It was in the woods. Closed on Mondays for maintenance. Friends fixed it so we could use it. I was on the phone telling Linda about it. Duke overheard me. *"Can Georgia come play, too?"*

Linda and Georgia came. Linda wanted to see the course We took a couple of clubs along. Duke, Georgia and Kitty rode in the golf cart with Linda. Duke and Duchess had their harness on. I knew Kitty would have to ride on Duke part of the time. Duchess, well she just liked the sparkles on her harness.

We hit four balls. Off they went. *"I'll get that one over there, Duchess,"* Yelled Tiger.

"Oop's. Duchess, it's too big for me to pick up with my teeth."

"You find them, Tiger. I'll pick them up."

"Look, Duke has one and so does Georgia," Tiger said.

They dropped the balls in the carts and we hit them again. Three went fairly straight and one off to the right into the woods. *"We'll get it,"* Duke yelled with Kitty holding on to his back for dear life. *"Do you see it Kitty?"*

"No, lots of fallen branches and stuff. Let me down, Duke, and I'll sniff it out under the leaves."

"OK, just be careful."

All of a sudden there was a great shuffling noise and out of the brush ran this large gray body, it was almost seventeen inches long and a tail longer than that. It raced onto the course. Duke and Kitty ran out of the woods, stood and watched it run across the fairway.

"Wow! Look at that," Tiger yelled. *"See it, Georgia?"*

"It's big. What is it?" she asked.

Duchess trotted up to them. *"Looks like a giant squirrel."*

"Look at that tail. Gotta be a mile long," Tiger said with awe.

I pulled up to where the dogs were standing, *"Should we chase it?"* Duchess asked

"No!"

"What did you say?" Linda asked.

"I was thinking, I'm glad the dogs aren't chasing it."

"What is it? A monster squirrel?" Tiger asked. "

"It's a Fox Squirrel. They're seldom seen around here, quite shy around people. There's a large wooded forest south of us, across from where we live. It may have come from there. They're rare to see because they like old growth forests."

Duke and Kitty came trotting up. *"See what Kitty flushed out of the bushes."*

"Were you scared, Kitty?" Georgia asked.

"No. It just startled me. Look. It's stopping up ahead, sitting up and talking."

"Can you understand it?"

"No, it's just fussing at us. We disturbed its hunt for nuts."

"I never saw a squirrel that big before. It's sitting there, making noise at us. The dogs won't chase it, I hope," Linda said.

"It's running off. Wow! See it jump and go up that big tree." Duchess said. *'That was exciting. OK there's a green, hit some balls to it.*

"Linda, time to hit balls again. Lets try for that green up there"

We landed four balls on the green. In a minute or two the animals landed as well. Linda and I pulled the carts up. The games had begun.

"What in the world are they doing?" asked Linda, her mouth hanging open. "My word we better stop them. They might hurt….."

I put my hand out to catch her from walking over. "No Linda they won't hurt anything."

"You can go first, Duchess, then Georgia," Duke said. *"When you get them close enough, Kitty and Tiger will knock them in."*

"Duke, look at Linda. She seems upset," Duchess said.

"I think she just can't figure what we're doing," Georgia explained. *"She'll be OK. She's smart. She'll catch on it's a game."*

"Yeah. I hope that's all she catches on to," Duchess said.

They hit the balls till all four were in the hole. Then they sat and looked at us. I heard…

"See how good we are? Didn't take us more than three or four strokes each," Tiger said.

We hit more balls. Then one rolled down a gentle slope toward a pond. Georgia was after it and skidded to a stop. *"What's that?"*

"Don't go any closer Georgia." I heard Kitty yell. She was standing a ways behind Georgia and looking past her down the slope. Tiger was slowly creeping closer to Kitty.

"Oh boy! Hurry everyone. It's another sea monster," Tiger called.

"Is it moving?" I asked. *"Don't disturb it."*

"What is it?" Linda asked.

We slowly crept closer, as did Duke and Duchess.

"Oh My God! It's a gator. We have to get Georgia away from it." Linda started to get out of the cart.

I reached over and grabbed her arm. "No, don't move, Linda. You don't want to scare it into action."

"But. My baby !"

"Duke, you and Duchess get her away."

"They understand you?" Linda asked.

"Typical commands," I said covering up the other telepathic exchange with some verbal commands..

"Duchess, be ready, you're closer. Georgia, listen now, you slowly back up on

your belly. If the gator moves, Duchess will grab you and pull you back. No quick movements. Duke, you help if you're needed.."

"If the gator moves, I'll distract her away from Georgia," Kitty said.

"Oh, boy. That leaves me to save you," Tiger moaned.

"Listen up you two. You won't do anything but run to me, Kitty, and you too Tiger. I have a golf club. I'll distract it if it comes this way."

"Linda take your club and be ready to hit the thing on the nose if it comes up the hill. It's not a large gator. It's only a little over four feet long. A young one. It was probably just dozing in the sun."

"Georgia, start backing up slowly. That's it. Kitty, you and Tiger back up, too. Slowly now."

"It' opened one eye," Duchess said.

"That's fine you all just keep coming back. Georgia, you're doing well, no quick movements."

"It shut it again," Duke said.

"Into the cart Kitty, you too Tiger." They stepped up and hopped onto the seat.

"Linda," I said. "See, here they come and the gator couldn't care less."

The dogs jumped into the carts and we drove off.

"Oh Georgia, I almost lost you," Linda said, holding and rocking Georgia.

"That'll teach you to stay away from gators, Georgia," Duchess said. *"They can gobble up small dogs and cats and even drag bigger dogs and little kids into the water. They drown them, before they eat them."* I looked over at her. *"Well, the big ones do for sure"*

"I never saw one before. Thanks Duchess. I know you and Duke would have saved me."

"We again escaped the jaws of disaster." Everyone stared at him *"Well, you know what I mean. Can we play some more?"* Tiger asked.

"I don't think so, Tiger. That really scared Linda and Georgia and…me, too," I said.

"OK, lets go home," Tiger said. *"Maybe we can get a treat for being so brave. Huh?"*

They all nodded. Even Georgia was into this nodding, *"All right, I think we can manage that."*

7
Angel?

It was a hot summer afternoon when Tiger walked down the dock to where Kitty was sitting on the bench. *"What'cha doing, Kitty?"*

"Talking to the angel?"

Tiger looked all around, *"What angel?"*

"The one sitting on the handrail. Right there in front of me."

"Uh-huh," grunted Tiger. He cocked his head and looked closely.

"Can you see it, Tiger?"

"No Kitty, I can't. Can it show itself to me?"

Kitty slowly shook her head. *"I'm afraid not. You can either see them or not."*

"What were you talking about?"

"Birds."

"Birds? Well OK. By the way whose angel is it?"

"I didn't ask. I think it may be ours. Birds are really interesting. I like them a lot, Tiger. And it's OK to talk to angels."

"Okay, I'll see you later," Tiger said. He backed away, eyes blinking in a futile effort to see the angel as he left the dock and ran off to find the others.

* * *

I was trimming some trees by the road, under the supervision of Duke and Duchess. Actually, they were very helpful. They would pick up the branches and drag them to the cart.

Tiger came racing down the drive, skidded to a stop and said, *"I think we have a problem."*

"What?" I asked, sitting on the cart to rest.

"Kitty."

"Kitty? Is she OK?" Duke quickly asked.

"Yeah, she's OK. I mean she's not in trouble. She's sitting on the dock."

"Well then, what's the problem?" Duchess asked.

"She's sort of acting strange," said Tiger, with a half bowed head.

"Strange?" I asked. *"What's she doing to make you say that, Tiger?"*

"Yeah," growled Duke. *"Kitty isn't strange."*

"I don't mean she's strange. She's....well, she's sitting on the bench talking to an angel."

We all sat there with our mouths open and stared at Tiger. Duchess recovered first. She shook her head and said, *"How do you know that, Tiger?"*

"I asked her what she was doing, and she told me she was talking to an angel. I looked real hard and couldn't see any angel. I asked if the angel would show itself to me. She said no. That you could either see them or not."

"You're telling this to us, verbatim?" I asked.

"Ver… what?" Tiger said.

"You're repeating the conversation with Kitty word for word."

"Yeah. I couldn't make this up."

Duke, looking a bit alarmed, stood and said, *"It, this angel, it won't hurt her will it? I mean shouldn't I go and chase it away?"*

"Na," Tiger responded. *"You can't chase what you can't see."*

"OK, that makes sense. What were they talking about? Did you ask?"

"Birds."

"BIRDS!" We all said at once.

I recovered first, smiled and nodded.

"What are you smiling at?" Duke asked.

"Well, if I had to guess at what Kitty would find to talk with an angel about, it would be birds."

"Yeah." He replied, *"I guess so. She really likes birds. They seem to like her, too."* He paused. *"Don't they say angels can fly? And you know birds fly, and she can talk to birds…"* He looked at me and cocked his head. *"It does seem to, sort of, fit together."*

"Hey, are you guys losing it too? I mean do you really think she can see an invisible angel and actually talk to it?" Tiger asked. He was standing there with a forceful manner. I mean, if a cat could have its paws on its hips, with its jaw thrust forth, he would have.

That did it. Duchess jumped right in. She stared at me and said, *"You remember in the woods when we found that lost girl?"*

"Yes." I replied.

"That girl was talking to her friend."

"What friend? There wasn't anyone else there," Tiger said

Duchess stared at me and went on, *"You said it wasn't unusual for small children to have invisible friends."*

"I did say that."

"Yeah, you did and you made me give the friend a ride on my back."

"What's all this?" Tiger and Duke asked in unison.

So we told them the story about the friend and how the child wouldn't walk away with us and leave it. So I said it could ride on Duchess, because obviously it wouldn't weigh anything.

"Yeah, she's leaving out the best part. When it got on my back it ruffled my

fur I felt it and," staring at me she said, *"you saw my fur being ruffled, I know you did."*

Caught. *"Yes Duchess, I did see that."*

"You guys never talked about it," Tiger said, accusingly.

"No, we didn't. I guess we thought it best forgotten."

"I sure can understand that. People would think you were both strange." Tiger looked at Duke and he nodded.

"OK, let's get back to this angel bit. Is Kitty strange? Are there angels?" Duke asked.

"No, I don't think Kitty is strange. She talks to birds. There are some who would think that was strange."

"What do you think it is?" Duchess asked.

"I think she's well… gifted. Yeah she's gifted," I said nodding. Oh boy, I thought. Now Tiger has us all doing it.

"What is this 'gifted'?" Tiger asked.

"It's when someone has a special ability, like with music or art or ease with languages. People say the person is gifted."

"That's OK, but you didn't answer about believing in angels," Duke said. *"Do you?"*

"Some people say there are angels and some say no."

"What do you say?" Duchess asked.

"Yeah, come on tell us," Tiger said, he paused *"I know you do."*

"Do what?" asked Duke.

"She believes in angels."

We all just looked at him. *"Well, Kitty heard her talking to them one night. She told me about it. It was bedtime and she forgot something she wanted to ask you. So she crept back upstairs to see if you were still awake and heard you talking to an angel. So she left."*

"OK," said Duchess, with her head cocked to one side looking at me. *"Did you see the angel you were talking to?"*

I looked at all three of them and finally said, *"No Duchess, I didn't see an angel. Kitty heard me saying an old bedtime prayer."*

"What was it?" Duke asked.

"Angel of God, my guardian dear, To whom God's love commits me here.

Ever this night, be at my side to light and guard, to rule and guide."

"Oh," Tiger said. *"That's nice. I can do that, too."*

"Yes, you can, Tiger. You all can."

"Seems Kitty, can just take it a step further, huh?" Duchess said.

"Can we all walk out to the dock and see if any of us can see the angel?" Duke asked.

"You don't believe me," Tiger muttered.

"No Tiger," Duchess said, as she rubbed her head up against him. *"We believe you, but maybe one of us could see it too."*

"OK," Tiger said, begrudgingly.

The big trees above our heads were filled with chattering black birds as we all walked to the dock. Kitty was on the bench, asleep. No angel in sight. *"I guess we'll have to try another day,"* Duke said.

8
Jumping Jack

"We're going to see a new room that friends added to their house. It's on the second floor overlooking the waterway, like ours, so come on." I said to the animals.

"All of us?" asked Tiger.

"Yes. You'll need leashes."

"We won't run away."

"Tiger, people have rules about animals being leashed. But...you could stay home," said Duchess.

"Nope, I'll come."

* * *

The house was a few miles north of us. We got out of the car and my friend Betty Jean was standing in her yard looking up at the roof of the neighboring house. It was a second story house with stairs going up to a porch on the second floor. It had a slate roof and dormer windows also roofed in slate.

"Will you look at that!" exclaimed Tiger.

"Yeah, a dog is on the roof. How'd he do that?" asked Duchess.

How indeed? I thought. "Bit of a problem, Betty Jean?"

"You bet. I don't know what to be the most frantic about. Hagen falling off that small dormer roof, my son Matthew falling off the ladder trying to rescue him, or them both damaging the roof slates."'

"How come they have that little window with it's own roof sticking out of the big roof?" Duchess asked looking up at me.

They all looked at me inquisitively. *"There is probably a small room above the second floor and they put a window in for light. It's called a dormer window. If it hadn't been there he would have slid all the way off the big roof."*

"That sure is a long ladder. He won't slip will he?" Kitty asked.

"He won't wiggle loose from your son will he Betty Jean?"

"No. Hagen is his dog. We're taking care of it while he's at school. I think Hagen is too scared to move much. See my son is holding him real tight." You could see him clutching the dog to his chest with one hand as he carefully backed down the ladder. It was a small white dog with brown/tan markings.

"That's quite something. We'll ask Hagen how he did it when he gets down," said Duke.

"I hope you're not admiring what looks like a stupid and dangerous stunt. Are you?" I asked Duke.

"Well," popped up Tiger, *"none of us could manage it,"* he said in an admiring voice.

"You sure couldn't," said Duchess. *"We almost had to get a ladder to get you out of that big tree you climbed up and then were afraid to back down."*

"Hey…" yelled Tiger.

"Enough, you two," I said.

Betty Jean's son and the dog came down the ladder and he handed Hagen to her. She hugged Hagen and scolded him at the same time. Then she set him on the ground, hugged her son and turning to me and said. "Come on in, I'll pour a glass of wine or even two, after that adventure. We can sit in our new room."

After she he put Hagen down. The sniff and greet started between the dogs. Tiger and Kitty as usual stood back and watched. They all then followed Betty Jean into the house and sat just inside the door to the new room.

* * *

"How'd you get on the roof?" asked Duke.

"More important, why?" said Duchess.

"I was chasing a squirrel. It went up the stairs, onto the porch railing and jumped to the roof. I just followed it," Hagen said.

"Yeah, then what happened," said Tiger with glee.

"Well, it ran along the roof and leaped to a limb in a nearby tree."

"Yeah," said Tiger shaking his head. *"They'll do that every time."*

"But why were you on that roof?" asked Duchess.

Hagen hung his head, looking very embarrassed. *"I slipped on the slates, slid off the house roof and landed on the roof of that little window below."*

"I bet you were really scared. I would have been," Kitty said.

Hagen looked at Kitty, nodded and replied, *"I thought I was going to fall all the way to the ground. It was a long-g-g way down."*

"What happened then?" asked Duke.

"I was afraid to bark too loud. I was afraid to move or do anything. Those slates were really slippery. I barked a little when I heard her calling me."

"And...." growled Duchess.

"I saw her in the yard, and then she looked up and just stared at me."

"Did she say anything?" asked Duke.

"No. She just pulled her phone out and I guess that's when she called her son. She did look kinda scared."

"I think she may have been afraid if she talked to you, you would get distracted and fall," Kitty said.

"I can believe that. She probably knows you're not too smart."

"Duchess, that's not nice," said Tiger.

"Maybe not, but it's true."

Duke looked like he would interrupt, but Duchess just glared at him and continued. *"Stop making excuses for him. You saw how far up that was. You'd have to be out of your mind to get on that roof, or not have a mind."*

"How did you do that again?" asked Tiger watching Duchess out of the corner of his eye.

"I jumped from the porch rail to the roof."

"You mean, almost straight up in the air. No running take off?" asked Duke.

"How could you do that?" asked Duchess sternly.

"It was kinda easy."

"Easy?" Duchess questioned.

"Yes, I'm a Jack Russell Terrier," he said with his head held high.

"So?"

"So, we're athletic and agile and can leap very high, easily five feet or more."

"From a standstill? I don't think so. No one can do that," Duchess said.

"We can," Tiger and Kitty said in unison.

"That's different, you're cats," Duchess said looking them over.

"But Duchess, look at him. He's about as tall as Tiger and he's skinny," said Duke.

"Skinny? I beg your pardon. I'm muscular, and fast and very intelligent."

"Well, you can leave that last part off," sneered Duchess.

Kitty said ,*"I think he's a very handsome dog. He may be little, but you have to admit that was quite a feat. But Hagen, it might be better if you sort of slowed down and looked next time. I mean being bold and curious is nice. But well, falling off a roof is.... "*

"Stupid." finished Duchess.

"Are you going to get a whippin'?" asked Tiger.

"Oh no. She wouldn't do that."

"What?" exclaimed Duchess.

"Well, maybe a small slap or pat on the butt accompanied by a firm, no. Are you whipped?"

"No," Duchess replied firmly. *"We talk about things and agree what we can and can't do."*

"Really, can you do that? Can you talk with her?"

"Yes, but you can't tell anyone."

* * *

After Betty Jean and I had a nice chat I got up to leave and walked over to where the animals were sitting. *"OK, what's going on over here?"* I asked

"Do you like the room?" asked Duke.

"Yeah, it sure is big. Almost twice as big as ours," said Tiger.

Another subject changer I thought. I squatted down and stroked Hagen's head. *"Well Hagen, did you have enough excitement for the day?"*

He looked up at me and then at the others and replied, *"More than."*

I smiled, *"I won't ask you how you got on the roof. Duchess and the others will tell me. But it might not be a good idea to try it again."*

Hagen turned his head from one side to the other looking at me rather quizzically.

"Will they really tell you?"

I rubbed his ears and replied, *"You bet."*

9
Fawn

"Can we go out in the woods again?" Tiger asked.

Duke chimed,*" Yeah, now that Kitty can ride on my back, she could go too."*

I looked at them sitting in a row, radiating anticipation, ears perked forward and twitching, eyes big. *"We can do that. We'll go tomorrow early in the morning,"* Since they got the harnesses I knew they were eager to try them out in the woods.

"Kitty, you OK with this?"

"Oh yes, I've never actually been in the woods. It'll be so exciting...." she paused. *"I'll be able to see much more, way up there, on Duke's back."*

"Good, tomorrow is a weekday, so we'll be mostly alone in the woods. We'll go right after breakfast."

* * *

We arrived, parked the car under some trees, and started walking along a trail off the road. Duke and Duchess both had harnesses on and I had leashes in my backpack. Kitty scrambled along beside Duke. The woods smelled fresh and pine scented there were birds in the trees singing and you could hear small animals scurrying in the underbrush

"Can we run on ahead through the trees?" Tiger asked, *"It's not over grown, we won't get lost."*

"Sure, I'll stay on the path."

"OK, jump up, Kitty," Duke said.

I'd mostly lost sight of them in the trees. They blended nicely. I'd catch an occasional flash of white from Kitty. Sometimes I'd see one or the other up ahead near the path looking back at me. After a while I rested on a stump and drank some water.

Tiger raced down the path toward me. *"Hurry come, you have to come now."*

"Is someone hurt?"

"A baby is caught, hurry. Follow me." Tiger turned and ran off down

the path. Suddenly he turned into the trees and waited. As I would draw near, he would turn and run on ahead. Finally he stopped. *"See over there, by those bushes."*

I could see Duke and Duchess sitting a bit back from the bushes and Kitty nearer the bushes. In front of Kitty was a small fawn about the size of a medium sized dog. She was light tan underneath and brown on the back with white speckles in her coat. She was trembling, her ears prickled up and eyes rolling in panic.

"See the baby?" Tiger whispered.

"She's very scared," Kitty said.

"Can you talk to her?"

"No she's not old enough to talk. She's caught in something and can't get loose."

I crept closer, trying not to scare the fawn, any more than she already was. It looked like she had a cord or a rope of some kind around her upper leg. It was holding her up and at an angle toward a large tree branch, where it was tied.

"What is it?" Duke asked.

"It's a snare to catch rabbits."

"You mean bunnies?" Kitty asked with concern in her voice. *"They catch bunnies?"*

"Yes, I'm afraid so."

"But that thing could kill them if it caught them around the neck," Duke said. *"That's not right."*

"Can you cut the fawn loose?" Duchess asked. *"I mean we can't leave it hanging there."*

"Yes, I have a knife and scissors." I carried a backpack with water, a small emergency box, snacks, binoculars and other things I thought we might need in the woods.

I took out the scissors. As I knelt closer, the Fawn's eyes got bigger and it struggled. *"Kitty, come closer and give it a rub and purr, see if that will calm it."*

One of the fawn's forelegs was held off the ground. It's hind legs were on the ground but the other foreleg barely touched. With one arm around the fawn I reached up and cut the cord, gently letting the fawn down. It collapsed on the ground.

"Can't it stand?" Tiger asked, *"is its leg hurt bad?"*

"Let me loosen the cord. It seems very tight." I could see it was a slipknot but with the fawn wiggling and its weight, it had pulled very tight around the leg.

"Move a bit, Kitty. Let me hold her in my lap and work on the cord. I'll have to slip the tip of the scissors under it and cut a bit at a time."

The fawn was trembling and trying to pull away. I no more than got the tip positioned under the cord and it would get pulled out.

"Could you talk to it if it wasn't a baby?" Duchess asked Kitty, looking nervously behind her, *"I mean if you can talk to birds, could you talk to deer?"*

Tiger looked quickly around. *"Oh, Oh. We have company."*

I looked up and about 30 feet behind Duchess was a doe. *"I think that may be its mother. Duchess, you, Duke and Tiger slowly come over here away from her path, and keep a good distance from us."*

"It won't hurt you, will it?" Duchess asked.

"I don't think so."

"Do you want me to move, too?" Kitty asked.

"No, I want you to try talking to her. Tell her we're trying to get the cord off the fawn's leg. Ask her if she can quiet the fawn down."

The doe made her way carefully toward us. I heard Kitty making noises and she moved closer to me. I continued to hold the fawn half in my lap, half on the ground. It was still squirming. The doe looked us over. She than put her head down and nudged the fawn. The fawn stopped moving. I put the tip of the scissors under the cord, cut it and saw it fall away.

"Is its leg OK?" Kitty asked.

"I felt the leg and it doesn't seem to be broken, but it's probably sore. Lets see if it can stand." I stood the fawn on her feet and held her steady. It was a bit wobbly. Than it took a few steps toward its mother and leaned on her while being nuzzled.

"OK," Tiger said with glee, *"we saved it…."* Suddenly his eyes got really big. *"Wow look at that."* He backed up closer to Duke and Duchess. The fur was standing straight up on the dog's back.

I turned and coming through the trees toward us was a large buck with a big rack of horns. It stopped looked at the dogs, pawed at the ground and tossed it's head *"That's its daddy,"* Kitty said. *"It wouldn't hurt us, but I think we should leave so they can take care of the baby."*

"I agree." I said hesitantly, I wasn't as sure as Kitty. I backed out right away.

As we were leaving, Tiger asked, *"Can we stay a while longer in the woods?"*

"Yes."

As we walked along Duchess asked, *"Kitty, were you talking with the deer?"*

"Sort of."

"What do you mean?"

"I talked to it like we talk to each other, and also like I talk with the birds. I didn't hear an answer but I sensed it understood me."

"Maybe the doe was shy," Duke said.

"I'll bet it was scared. If you didn't have her baby she wouldn't come near you," Tiger said.

"I think Tiger's right. If I had a puppy I'd stay close by to see what you'd do. When it was cut down and Kitty was nuzzling it, rubbing and purring, I think she realized it was OK to come closer," Duchess added.

"Yeah, and then you started talking to it," Duke said.

"Well, we saved that one," Tiger said nodding. *" But maybe there are more of those things out here. Shouldn't we look?"*

"I think so," Duke replied looking at me. *"Is that OK?"*

"Yes, but be very watchful and careful where you run, and don't lose me" They took off.

Duke ran toward me. *"Come, we found more."* As I approached Duke said, *"See there's one. Kitty found another and Tiger and Duchess found two more."* I cut them all down and wound the rope into a small bundle all the while swearing quietly to myself about the cruelty. We were in the Guana Wildlife Reserve and I was sure such behavior was not approved here. We were in the far north end, away from most of the high traffic paths.

Duchess came running up just as I was finishing with the rope. *"Kitty's crying and really upset, she found a bunny and it's not in good shape."*

I walked through the trees after Duchess. Kitty was sitting near a rabbit hanging in the air by it hind foot. I ran to it. It was still alive but very weak from struggling to get loose from the rope holding it up off the ground. I quickly took the weight off the foot and held the rabbit tight to my chest and cut the rope from the tree.

It was very weak and I was able to cuddle it in my lap while I cut the rope from its foot. Kitty rubbed on it and purred loudly. It didn't look like its leg was broken. *"It may need some water to drink."* Kitty whispered.

I poured a bit into a small bowl from my backpack and held it under the rabbit's mouth. It tentatively drank. Looked at me and at Kitty, then drank more. I set the rabbit on the ground and put the bowl beside it.

"I guess all that struggle got it thirsty, huh?" asked Tiger.

"Can I keep the bunny? I found it." Kitty asked very quietly.

"No Kitty, they are not house pets. This one has a home out here in the woods and maybe even a family."

"I know, but it didn't hurt to ask, did it?

"No you can always ask. I like bunnies too. Let's walk back a ways and give it a chance to hop away."

The rabbit watched us leave, looked all around, then hopped into the bushes and was gone. *"Time to go home."*

"Can you call someone and see if we can stop this snare stuff?" asked Duchess.

"Soon as we get home."

"Thank you, I'm glad we came today." Kitty said and the others echoed, *"Yes, me too!"*

10
Cat in the Tree

Duke and Duchess were helping me pick up the yard down toward the road. According to Duchess, Tiger was along for the ride. He couldn't carry anything but was great at pointing out stuff we should get, then when the cart was full he'd hop on and ride to the road. I called it supervision. Duchess called it disgusting. I sat on the cart to rest when Duke asked, *"Where's Kitty?"*

"She said she would pick up our toys and stuff and then come join us," replied Tiger.

The animals were good about helping out. They were very creative in finding ways to assist. They had proven their usefulness when Tom built the stairs.

"I'm not comfortable. Something doesn't feel right," Duke said. *"Tiger, let's go see where she is."*

"What do ya think, Duchess?" She was following them across the yard with her eyes.

"I think I'd like some water, how about you?"

"Yeah." As we were walking to the house, Tiger came racing toward us.

"Hurry, Kitty's up a tree, a big tree."

We followed Tiger to the north side of the house where there was a very large oak tree. Duke was sitting on the ground looking up. There was a large split up the side of the trunk that opened nearer the top, before the tree divided into larger branches. It looked like it might contain a sort of hollow. Kitty had her head poking out of it peeking down at us.

A tall palmetto bush had grown up along side the tree. The tall fronds reached almost to the hollow that Kitty was in. It was a good fifteen feet high.

"What's going on, Duke?"

"Kitty said she was running for her life and hid up there."

"Kitty," I called.

She peeked out a bit further, *"Is it gone?"*

"Is what gone?" asked Tiger, looking all around. *"I don't see nothing."*

"It was that big bird. You called it an eagle. The one that tried to grab me off the boat when we were at that island."

"Tell me what happened, Kitty," I said.

"I finished picking our toys up and came over this way to do my business under the bushes. I heard the leaves and small twigs, like cracking. I looked up and this huge bird was coming thru the trees across the yard toward me. I raced behind these tall fronds below me and that stopped it for a moment," she paused.

"Yeah, Yeah, then what?" Tiger urged.

"It landed and stalked toward the tree."

"Oh Boy! Where is it? I'll kill it." stormed Duke, pacing and looking all about.

"Then what? Then what? What did ya do?" Tiger asked urgently. You could almost hear him saying, "what an adventure".

"I scooted up this tree through the large crack and ducked into this deep hollow. I don't think it saw me do it, cause the palmetto fronds were sort of hiding me and it was looking on the ground and in the bushes."

We all looked where Kitty said she had ducked into the crack and climbed up. The fronds would have hidden her nicely if the eagle was looking on the ground.

"It was a very large bird."

"Bet you didn't talk to it, did you?" Tiger quipped.

"Tiger," said Duchess, moving over toward him.

He scooted around behind me, *"I didn't mean anything, Duchess. You know she's always talking to the birds."*

"Not the ones who are hunting her to eat, dummy."

"I didn't mean anything," he repeated as he hung his head.

"I thought about it, Tiger, but then it would know where I was hiding. The eagle might have tried to reach in and snatch me out."

"You're right, Kitty. Then what happened?" I asked.

"Duke and Tiger came racing across the yard and he flew away."

I thought about what happened. We'd been working in the front yard toward the road for quite sometime. Often large osprey and even eagles landed on the tall branches of the oaks on the waterfront and watch the water for fish. Kitty's bright color and the absence of the dogs and me must have been very tempting. Kitty was a very small calico, just a tasty size to snatch for lunch.

I looked carefully up at all the trees and no one was about. *"OK, Kitty come on down. It's over."*

"Down?" I heard.

"Yeah, down. Climb down."

Kitty looked and looked and finally said,*" I don't think so."*

"What do you mean? You don't think so. You can't stay up there," Duchess

said.

"I don't know how to come down."

"Same way you got up there, in reverse," Duchess' patience was waning.

"I was so scared, I just climbed as fast I could. I don't remember much."

"Oh for heaven's sakes. Tiger, you climb up there and show her how to climb down," Duchess commanded.

"Me?" He looked up, *"That's a long way up there."*

"Well, it isn't going to get shorter with you standing here looking at it. Go on."

"I don't know, Duke?"

Duke looked up the tree at Kitty. Then he looked at Tiger and finally at me. *"Do you suppose you could get a long ladder and get her?"*

"No, I don't suppose I can. Duke, She has to learn sometime how to climb both up and down trees. It's part of survival. Especially for cats," I replied. *"Kitty, you just turn around and come down backward. No rush, take your time."*

Duke looked at Tiger. *"You can show her. I know you know how to climb trees. Please."*

I walked over and sat down on the porch steps and watched. Duchess sat with me. She whispered, *"She'll be OK, won't she?"*

I put my arm around her and rubbed her ears. *"Yes, she'll be OK. So will Tiger. We'll just sit here and watch them figure it out."*

As Tiger went up the tree to the side of the large split, a bit above Kitty, he told her to hang out a bit and watch him climb.

"OK, you got it?"

"Yes, I think so. Now go down and let me watch." When he got about three feet from the ground he turned and jumped.

"Now it's your turn. No big deal, as easy as swimming was."

We watched every step she took coming down. She was a bit hesitant at spots, then she spun and jumped. Duke was all over her licking and pushing with his nose and asking *"You OK, you sure?"*

She pushed him back with her nose. Sat, washed, and said *"Yeah, a piece of cake, as Tiger says."*

"What do you think we should do about this?" asked Duchess.

"Do about what?" Tiger asked. *"She's out of the tree, it's over. We need to get back to work."*

"Over for now. What about the next time?" Duchess said with a sneer. *"That eagle isn't going to forget she lives here. Maybe he'd even like a piece of you."*

"Come on Duchess, that's mean. It won't go after me, I don't think. I'm too big." He paused, *"Aren't I?"*

"I don't know Tiger. They pull some large fish out of the water and, carry them off. I saw one even half fly and half swim one to shore. The fish was so big and heavy the bird couldn't fly with it. So it just dragged the fish to shore and finished it off there. You saw that too, didn't you?" She said turning to me.

"Yes, I did. It's just possible the eagle could snatch Tiger."

"Hey, come on. You know Kitty likes to sit on the dock, even sleeps in the sun. She shouldn't do that anymore, should she?" Tiger was sounding really agitated. *"Duke, what do you think?"*

"Gotta hide Kitty," He said. *"Maybe you, too."*

"OK, that's enough of this scary stuff," I said. *"We're not going to hide Kitty or Tiger. But Kitty, it may be a good idea if you don't sleep on the dock in the sun, without one of us with you. We have enough tree shelter that if one flies thru you will hear it just like this time. For the next day or so, be careful and have one of us in the yard with you."*

"I'll watch you, Kitty. I won't let anyone get you," Duke pronounced. *"You either, Tiger."*

"OK, let's get a move on." I said as I got up and walked off. *"We still have work to do,"*

11
Shots

"You have to go to the veterinarian's tomorrow to get your yearly shots"

"Shots!" exclaimed Tiger. *"I don't think so. That's scary and it hurts. Uh-uh, not me,"* he said shaking his head.

"Tiger, what's going on? It's only a little stick," Duchess said. *"Duke and I had to get those a lot."*

"Good, you can get mine for me. I'm not doing that. Uh-uh, no way," he said backing away from us.

"You know, you can't live here with us if you don't get your shots. We'll have to take you back where we got you," Duchess said, sounding sorrowful. *"Be a shame, you being a talking cat and all."*

Tiger pulled up tall, sat, and looked at all of us. *"You mean you would let them take me and put me down?"*

"Sure wouldn't want to. We've gotten used to having you around. But those are the rules. Gotta have your shots to keep you from catching something. You might get sick and worse, you might pass it on to us. Isn't that right?" She asked looking at me.

I was speechless, listening to this interchange. I didn't expect Tiger to be so vehement nor Duchess so blunt.

"I'm not going to take you back. But you are going to have your shots," I said firmly.

"Tiger," Kitty said in a very soft voice. *"I would be very sad if you got sick, and, and couldn't get well."*

"Me too, Tiger," Duke said, walking over and rubbing his nose in Tiger's fur.

I left the room.

* * *

We arrived at the vets the next day at 10 AM with the dogs, in their harnesses and everyone on leashes. They decided the cat should ride on their backs instead of me lugging cages. The cats sat in one chair and the dogs on the floor in front of them. I walked to the counter to check us in and I heard … "Excuse me Miss, is it safe to let them sit there without you holding their leashes?"

I looked back. Adjacent to the row of seats the dogs were in, was a woman sitting on the edge of her chair, clutching the leashes of two small white fuzzy dogs in the chair next to her..

"They won't move," I replied.

"Well, Tiger would if we let him," I heard Duchess say.

"Shush, Duchess. Don't be picking on him now," Kitty said.

The woman said, "Are you sure they won't attack my Fluffy and Honey Bee?"

Tiger exclaimed, *"She's kidding isn't she? Your names aren't really Fluffy and Honey Bee, are they?"*

"We're Maltese, and when she bought us she took one look and said 'Oh my Fluffy Honey Bees.'

"Which is which?" asked Tiger.

They told him, than asked, *"What are your names?"*

Tiger told them.

Fluffy, then asked Kitty how come she didn't have a name? *"I mean Kitty is what you are, it's not a name."*

"Yes it is. I was asked what I wanted to be called and I picked Kitty."

"Oh," said Honey Bee, holding his head down. *"I wish I could pick my own name."*

"What would you pick?" asked Duke.

"Killer."

"Killer?" questioned Duke with his head cocked to one side.

"Yeah," Honey Bee responded. Sticking his chest out, showing his teeth and growling. *"See."*

Duke started to respond and Kitty interrupted, *"That's scary, I'll call you Killer, you act like a Killer."*

"Thank you…"

The woman snatched up Honey Bee, "See, they're scaring my darling. You better get over there so Honey Bee will behave."

When I sat down next to the cats, I looked at Honey Bee and said, *"Hi Killer."*

He sat up, threw out his chest again and said, *"Yes I am and thank you."*

Just then the door slammed open and two rambunctious young boxers rushed in dragging their owner behind. The owner was a young woman and she was yelling at them to stay, pulling for all she was worth on the leashes.

The boxers stopped moving but were leaning forward. I heard, *"Hey, look at that, cats."*

"Yeah, but look at the police dogs in front of them."

"You're right but there's nothing in front of those sissy little white dogs. Let's get um."

As they pulled toward Honey Bee and Fluffy, the woman grabbed them up and moved quickly over into the chairs near Duke, all the while yelling, *"Control your dogs."*

Duke and Duchess stood up and moved in front of Honey Bee and Fluffy. Duke said, *"You two, sit down and shut up. We don't want to hear another growl out of either of you."* They towered over the boxers.

"These are our friends," said Duchess. *"You understand? Now sit."*

The boxers quickly sat in front of the chairs by the door. Duke and Duchess resumed their place in front of the cats but closer to Honey Bee, never taking their eyes off the boxers.

The door opened and an attendant called for Honey Bee and Fluffy. As they went by, in the arms of their owner, she turned looking at the shepherds and said to me, "Thank you."

I also heard, *"Thank you,"* from Honey Bee and Fluffy.

* * *

Soon it was our turn. Tiger and Kitty hopped up on Duke and Duchess's backs and we all trooped in. The Veterinarian, Jack, who was my next-door neighbor, was not surprised to see my cats riding on the backs of the dogs. He had seen them outside practicing in the yard

"Who's first?" He asked me.

Duchess said, *"Duke and I will go first."* They both stood, the cats hopped down next to me and the dogs walked over to Jack. As I quickly said, "They are."

"Lets go in back," he said.

"I'm not going in back with him. I'm telling you now," Tiger said.

"Ah Jack, can you give them their shots out here? I think they'll be less scared if they stay together."

"Oops," said Duke. *"I forgot. I won't leave you, Tiger,"* He then stood next to Jack.

"OK Duke. Two sticks." He looked at me and said, "But I have to walk them back to get weighed."

"No problem." I looked down at Tiger. "I'll carry Kitty, and Tiger."

Tiger's eyes were big as he watched Duke and Duchess, standing very still while they got their shots. He eyed the syringe and the needles carefully. *"The needles aren't really too big are they?"* He asked, looking up at me.

"No, they use the smallest they can. But it will sting, so don't jump and pull it out or it'll have to be done again."

"Twice, no way. I'll be still."

"Did it hurt bad, Duke?" asked Kitty in a small voice.

"No Kitty, like she said, just a sting." He looked at her. *"Kitty, you'll be OK. She'll hold you. Won't you?"* he said, looking up at me.

"Who wants to go first?" I asked.

"You decide, Tiger." Kitty said.

"OK," he paused and looked at Kitty. *"I'll go first."*

I lifted him up on the table and put my arms around him. When the vet's helper came over, I said, "I'll hold him. He won't move."

"Better not," Duchess said.

"I won't move, I promise."

He did twinge a bit on the first one and pushed a bit tighter in my arms but stayed still.

"Good job Tiger, all over." I set him down.

"OK Kitty, you're next," I said out loud as I picked her up.

She eyed the needle, *"They look big to me."*

"That's cause you're little, Kitty," said Duke pushing up against me.

Jack looked carefully at Duke. "He seems very protective of Kitty."

"He'll be fine, Jack. Kitty belongs to him. I mean he practically raised her."

When it was over, Kitty quickly licked, the spot with her tongue and shook her head, like something tasted bad.

"OK, lets all go get weighed," Jack said.

I watched as each one hopped up on the scale. I was relieved it was over.

"Those are nice animals. They behave well," said Jack.

"Yeah, except for the big cat," sneered Duchess. *"Hop up Tiger, you did well I'm just kidding."*

"Kitty, I'm so proud of you. You too, Tiger," Duke said. *"It's over till next year."*

"Till when?" Tiger yelled.

"Forget it, Tiger. A year is a long ways off," said Duchess as she strutted out to the car.

12
Dog Park

"You sure Georgia will be there?" Duke asked for the third time, as he panted excitedly. We were in the car heading south to St. Augustine Beach to a new dog park that Linda told us about.

"Enough already," Duchess said. *"You've asked three times. We know you're smitten, but really..."*

"Duke, have you ever been to a dog park before?" asked Kitty.

I knew she was trying to soothe Duke.

"No, I don't even know what one is."

Before Duchess could jump on him again, I said, *"It's just a large fenced-in piece of land where dogs can run free, off leash with other dogs"*

"OK, OK, I remember you told us that."

"You sure it will be OK for Kitty and me to be there?" Tiger asked.

Duchess replied. *"I don't know why not. You were at the parade with us. No one will bother you with us around."*

"Yeah, but those dogs were on leash."

Duchess pulled back from the window. *"Tiger" She* said, with clenched teeth, if such a thing were possible with a dog.

"OK, OK. We'll wait and see. I can always sit in the car."

"Yes, you can."

I heard her grumble something else in her throat. I quickly said, *"It's right up ahead. See, there are only a few cars. We should be OK."*

* * *

The cats were riding on the dogs, who were in their harnesses. There was a bench near the entrance gate under some trees. Linda was sitting there and waved us over.

"Hi," I said, as I sat down.. *"Hop up on the bench Kitty and Tiger, you can sit between us."*

"Oh look here comes Georgia," Duke said, with glee.

Georgia slid to a stop in front of Duke and Duchess. After some mutual sniffing, Georgia asked, *"Still no sniffing, Tiger, Kitty?"*

"If you put your nose up I'll give it a gentle pat," Kitty said.

"No claws," replied Georgia, as she put her nose up to their paws and got a gentle pat.

"We wouldn't do that, you're our friend," Tiger said.

"What are they doing?" Linda exclaimed.

"It's a substitute for sniffing. The cats don't do sniffing," I replied.

"What?"

Oh. Oh. I thought. Careful now. "I don't know, it just seems to me they have a need to touch body parts."

"That's a good one," Duchess sneered.

"Hey, there's Mr. Dane," Kitty cried as the Great Dane dashed up.

"Yeah, saved by a Dane," Tiger said.

"Hi," said the Dane, *"lets all go for a run."* He gave a playful bark.

"Will you guys be OK here?" Duchess asked.

"I'll keep an eye out and she can call us," Duke said nodding at me.

"Go ahead, run, have fun," I said.

"Will you look at that, what a lovely sight, beautiful dogs," Linda said, as she stroked Tiger's fur.

" Maybe so," Tiger said. *"But we get the pats."* He began to purr.

"Come on meet the others. Maybe we can have races," the Dane said.

Beside our three and the Dane, there was a Bichon Frise — a lovely white ball of fluff — a Yellow Lab, and a Springer Spaniel. After they all milled about smelling each other, they lined up near one end of the park and raced to the other end.

"Look, The Bichon and the Spaniel came in last, but not by much." I said out loud.

"I think that's because their legs are shorter," Kitty said looking up at me.

"Yeah," Tiger said. *"But that little white one is running circles around them when they're just playing tag."*

"You know it looks like they're playing some kind of game," Linda said.

"I wouldn't be surprised. Dogs are very smart," I replied.

"And we aren't?" piped up Tiger.

I reached over and ruffled his fur and said, *"Tiger you and Kitty are the smartest cats I have."*

"Very funny, since I don't see any others at home."

"Tiger, stop being so sensitive, or as Duchess would say 'enough already,'" Kitty said. *"You are a big, strong, beautiful cat. There is no reason for you to sound so insecure."*

He looked at Kitty really hard, bowed his head and said, *"OK Kitty, and thank you."*

* * *

The owners of the other four dogs were on benches near ours and reading books or papers. Duchess, Duke, the Dane and Georgia came over to us. Duchess said, *"The others would like to meet the cats. OK?"*

"Yes."

"OK," she said, *"Come on slowly, and no sniffing."*

They all came and Georgia said, *"No sniffing. Just put your nose up and they will give you a gentle pat."*

The Dane did so right away. The others said they would wait till they knew them better. I took that to mean they had some experience with claws.

"I want you and Tiger to meet Bejee. She lives with me," The Dane said, nudging the Bichon with his nose.

"What kind of name is that?" Tiger asked the Bichon.

"Short for Bichon." He hesitated then added, nodding at the Dane, *"His name is Dane. Our owners are sort of literal and very busy people. So calling us what we are seemed simplest to them."* He looked at Kitty and asked if that happened to her, too. Kitty told him no, she picked her own name.

Tiger asked, *"What kind of word is literal?"*

"I don't know I just heard it at home as a name one called the other."

The Lab told them her name was Sammie, and the Spaniel was Boots because the fur on her feet was brown against the white legs.

I reminded myself to ask Linda about the owners of the Dane and Bichon. Could they be literal like Bejee said?

They flopped on the ground around us panting and grooming when I heard. *"Can she really talk with you and understand you?"*

"Yeah," Duke said.

"She can talk with me and Dane, too," Georgia added.

"Can you tell her what you want to eat?" Sammie asked.

"What do you mean?" Duchess asked.

"I mean, could you tell her you don't want any more dry food, you want steak?"

"Well yeah, we could," Duchess replied

"A lot of good it would do," Tiger responded. *"I mean asking and getting are two different things."* He looked back at me and I smiled.

"Do you get canned meat ever?" Sammie asked.

"Yeah, she puts a little in our dried food. Mixes it up you know," Duke said. *"Same with the cats but it's served separate for them. Duchess and I like it mixed in."*

Dane and Bejee said. *"We had a problem till Georgia told us how she trained Linda."*

"How'd you do that Georgia?" Sammie asked.

"Guilt."

"Guilt?"

"Yeah, I quit eating and she got so worried she'd try tempting me with a bit

of meat in my food."

"It worked?"

"Yeah. If she gives me something new I take a morsel in my mouth very briefly. If I don't like it I politely and gently open my mouth and drop it on the floor. Then I look pitiful. It works. I don't get any more of that. You have to be careful and not carry it too far. Humans aren't really dumb you know."

"You eat OK, Boots?"

"Yeah, the people they got me from told them how to feed me. I eat like you guys do."

"Dane, Bejee come on, time to go home." Their owners called.

"OK guys, I hope we can meet again some time," they both said.

Linda said, "Well that was fun and the dogs behaved well toward the cats. We'll all have to come again. Georgia seems to really like running with Duke, Duchess and the rest."

"You don't know the half of it," Tiger said.

"Linda,what do the owners of the Dane and Bichon do for a living?" I asked.

"College professors, in math and science. They don't talk much but are really nice. Kind of literal," she grinned. "They named their dogs Bichon (Bejee) and Dane. Sort of like you, with Kitty."

13
Squirrels

"Excuse me." I heard Tiger say. I looked up from my book. He and Duke were sitting in the doorway.

"Hi Tiger, how can I help you?"

"We have a problem with Kitty and some squirrels."

"Squirrels? What's the problem? Duke, is Kitty OK?"

"Well she is so far. We have Duchess watching her?" Duke replied. *"You need to come see and help us."*

"You'd better tell her what's going on first, before she goes out there," Tiger warned. *"She'll have to have her thinking straight and ready."*

"My thinking what? Duke, what in the world is going on?"

"It's not in the world, it's in the yard and, and, you have to fix it."

"Just tell her Duke. Stop stammering," Tiger said with some impatience.

"OK, OK. Kitty found some baby squirrels and started playing with them, like hide and seek."

"They were over in the corner under that tree and in those bushes. You have to go almost into the bushes to see them," added Tiger.

"Yeah, we wouldn't have known but we saw flashes of white and yellow and went to see what it was. There was Kitty with baby squirrels. I think they fell out of the tree early. I mean they weren't hurt, but we don't see them that young."

"How big are they?" I asked.

Tiger and Duke looked at each other. Finally Duke said, *"Like Kitty or smaller when we got her."*

"They have fur and a tail and can scamper around real well," Tiger said nodding at Duke. *"Isn't that right?"*

Duke nodded, "Yes."

A large motor cruiser was moving much too fast down the waterway throwing large wakes against the bulkhead. We stopped and watched it. Then I said, *"I see."* I thought a minute, *"They're probably still nursing.."*

"They what?" asked Duke.

"Nursing" replied Tiger. *"Babies have to suck on their mothers till they are big enough to eat real food."*

"Yeah? Where did you learn that?"

"On TV."

"I didn't see that."

"That's cause you fell asleep."

"OK, I guess I'd better go see." I said standing up.

"No wait. That's not all." Duke rushed on. *"The mother squirrel came and must have told the babies they had to get on the other side of the fence."*

"There was a lot of chattering going on and the babies scampered thru the holes in the fence to the other side. They squeezed through easily," Tiger said. *"Kitty was really upset, she was making noises at the mother squirrel. It seemed that the mother squirrel wasn't happy at all with Kitty and the whole play thing,"*

"That's when we walked up, really quiet like. We asked Kitty what was going on. We tried to stay out of sight of the mother squirrel. I don't think we were successful." Duke gave me a beseeching look. *"The mother squirrel got louder and louder, and Kitty said something and backed away toward us."*

"She asked us all to move back, that we were really scaring the mother. I could see that, but the mother squirrel was still noisy even when we backed up," Tiger said. *"Kitty told us, she was going to tell the mother squirrel that the rest of us would never go near that corner. That it would be OK for the babies to play with her in the bushes under the tree."*

"What was Duchess doing all this time?" I asked.

"Just watching," replied Duke. *"Anyhow Kitty went back up to the fence and there was a lot more chatter and noise from Kitty. I don't think she was convincing the mother squirrel of anything. Then she came back to us again and said the mother squirrel would not let the babies on this side of the fence at all. So she was going to go over the fence and play with them over there. She said she told the mother that. The mother wasn't too happy at all. But Kitty said she would get used to it."*

"Wow! That did it. Duchess spoke up and said Kitty was not going over any fence for any reason." Tiger said. *"Duchess said that Kitty knew the rules and was not going to break them for any squirrel play time. She was really firm, scared me."*

"So here we are," said Duke. *"It's a stand off. Kitty and Duchess agreed to stay still till we got you. Now what?"*

Now what indeed, I thought. I bet this mother was one of the many squirrels that the dogs and Tiger were forever chasing in the yard. They never caught and killed any but only because I told them they couldn't. No reason for the mother squirrel to trust that bunch.

* * *

I walked across the yard and squatted down by Kitty. Duchess backed up and sat with Duke and Tiger.

"You like the baby squirrels, Kitty?"

"Oh yes, they're so cute and they're smaller than me and very playful. I even touched one's tail. I want to go over the fence and play with them, but Duchess said no."

"I know. Tiger and Duke told me. They said the mother squirrel wasn't too happy about you coming over the fence. Is that true?"

"Yes, but she'll get over it and see I mean no harm. I won't hurt them. I told her so."

"Kitty, I don't think she wants you anywhere near the babies."

"Why? I said I wouldn't hurt them. I told her over and over."

"Kitty, dogs and even cats are very dangerous to squirrels. They catch them, kill them, and even eat them."

"Not me. Not us."

"How would those babies learn the difference…"

"I'd….."

"No, Kitty. You were going to say you would teach them. It doesn't work that way. Their mother will teach them who to stay away from, who to run from. Like cats, dogs, people and anything bigger than them or different from them. It was an accident that the wind blew the nest from the tree. I'm not sure they can even climb a tree yet."

"What will they do? Just sit under those bushes? What if something else gets them?"

"I'll get a box and put a towel in it and put it deep in those bushes. No one is home in that house next door. So they may be OK. We'll hide them from sight so an osprey or eagle doesn't get them. Now, I want you to tell the mother squirrel you're sorry you scared her. Also tell her I'll put a box with a towel in it for the babies. Then tell her we will all stay far away from that corner till the babies are big enough to get up a tree, or until she carries them off."

"What do you mean carries them off?"

"The mother squirrel gets a secure grip with her mouth on a baby's throat on the underside of the neck. Immediately, the baby automatically curls up around the mother's head. The mother makes sure her eyes are not covered and then with the baby tucked tightly around her head, she can carry it to a new nest. She can even jump a gap in the trees of five feet or more with a baby in her mouth.

"Do you understand?"

"Yes." Kitty said in a hushed voice… her eyes not meeting mine.

"Also Kitty, you know you can't leave the yard. None of you can."

"Yes."

"Good now tell the mother squirrel what I'm going to do."

* * *

Near evening I heard Kitty say, *"Can I sit in the chair with you?"*

"Yes." I hadn't seen her or any of the others since the squirrel episode. I was a bit concerned about Kitty. She had never been disobedient, stubborn or argumentative with the others. In fact she seemed to understand the rules and even helped Tiger understand, when he got rambunctious. I had never had to correct her before. She sat quietly. I waited.

"Are you mad at me?" She asked, in a soft voice.

"No, I'm not mad."

"I wasn't very nice to Tiger and Duke when they reminded me not to go over the fence. Then when I said I was going anyhow, Duchess was all over me. Scary."

"Uh-huh," I muttered.

"I just finished apologizing to all of them," she paused. *"That was hard. I want to say, I'm sorry, to you too."*

"OK." I paused and reached over and rubbed behind her ears. *"That was brave of you. It's hard to admit when you're wrong."*

"Will any of you trust me again?" I noticed the others laying in the doorway listening. Duke's ears were alert and almost stiffened to attention.

"Yes, Kitty. We know you have a kind heart and always want to do the right thing. We love you. You're a good Kitty."

"Oh." she looked up, stood, took a turn or two around. Curled up and went to sleep.

14
Birthday Present

In the evening I settled into the big chair in front of the TV with my book. It was quiet.. I assumed the animals were out for an after dinner run. I knew they'd be in shortly to watch TV. They were allowed to watch it only when I did.

I heard nails clicking on the stairs. I looked up. All four lined up in front of me.

"We would like to talk about a birthday present," said Duke.

"OK. Whose birthday?" I asked.

"Yours," Duchess said.

"Yeah. You said it was after the cold time. Isn't that soon?" Tiger asked. *"When it starts to get warm again, just before the hot time."*

"Yes, I said that. It's coming in two weeks."

"Well, we want to get you a present." Kitty jumped up on the chair next to me. She'd gotten into the habit of sitting with me when I read and watched TV in the evening.

"That's very nice. What did you have in mind and how will you pay for it?"

"We want to give you a picture of us." Tiger looked around for approval; they all nodded.

"Yes, and we want to have it in a beautiful frame..." Duchess paused, *"a really nice frame. It could even have sparkles on it."*

"No sparkles, Duke said. *"But we want it to be lovely for you,"*

"Well... yes," Tiger said, *"but it has to be lovely for us too. I mean... to show people how beautiful we are. I mean... to show them what good taste you have, I mean..."*

"Tiger," Kitty said, *"I think she knows what we mean. You don't have to worry, she doesn't do tacky."*

I looked down at Kitty. *"Tacky? Where did you hear that one?"*

"When Linda was telling you about her new neighbor's golf outfit. She said it was not only outrageous but tacky. Like she'd just gotten off the boat." She

looked up at me with a puzzled expression. *"Does that mean if she wore it on a boat it would be OK, but not for golf."*

I rolled my eyes and said, *"No Kitty, that's not what she meant, but it doesn't matter. I get the message. You want a very nice frame. Now, about paying for it."*

"We talked about that," Duchess said *"We'll work for the money."*

"Yes," Duke said. *"We'll pick up all the palm fronds and branches and put them in one or two places along the drive. Then all you have to do is put them in the cart to take to the road."*

"Yeah, that way you won't have to walk all over the place getting them," Duchess added.

"Kitty and I could pick up the smaller pieces," Tiger volunteered.

"And," I said.

They all looked at each other, with their faces scrunched up, like they were thinking hard.

"I know," Kitty said. *"We could pick up our toys, the balls and stuff, and put them in the basket before we go to bed."*

"Yeah, and before company comes," Duke said.

I was having a hard time keeping a straight face. They were trying so hard and were so serious. *"OK, that'll work."*

"Can we start on the picture taking tomorrow?" Tiger asked

"Yes, we can start tomorrow."

"Good, I'd like my red collar on. But you have to be sure the sparkles show."

"I'll do my best, Duchess."

* * *

The next morning I put Duchess's collar on her and gave each one a light brushing.

"OK let's go. We can start on the dock. Kitty and I can sit on the rail above Duke and Duchess," Tiger said.

"Why above us?" Duchess asked.

"Well, that way we can get the waterway in the picture, too."

"I'm not so sure…" started Duchess

I interrupted. *"It's OK Duchess, we can start that way. We'll take a bunch of pictures and you guys can choose."*

"OK, now how about in a line, side by side. Tallest to shortest," Duke

suggested.

They all lined up while I fiddled with the camera. It was a Lumix and I didn't use it much so I had to go over the settings again. I switched from scenic, to portrait, then settled on simple and worked with the zoom. It took a while.

"It's not broken, is it?" Tiger asked.

I had to laugh he had about as much patience as I have. *"No it is just taking me a few minutes to get the right settings. OK let's go."* I said with a smile. I shot several of them in a line.

Than Kitty asked, *"Can we do one with me standing next to Duke?"*

They shifted positions several times. Side by side with Tiger and Kitty in front of Duke and Duchess and several other combinations.

"Does my collar show in the pictures?" asked Duchess.

It was a digital camera, so I ran through a few of the shots. *"Not too well, Duchess."*

"Oh." It was a quiet, soft spoken, 'Oh'

Tiger looked at her and than at me. *"Maybe you're too far away to see it. Can we do some up closer?"*

Duchess looked at Tiger and perked right up. *"Could we?"* She asked.

I moved a bit and set the zoom in for a closer shot and arranged Duchess collar so it would show a bit. The cats were in front and Duchess cocked her head up and to the side. There was the collar in all it's glory. I shot several like that and Tiger did some cute poses in front and to the side of Duchess. She was so concerned about the collar she didn't pay any attention to him.

"OK, now can we do some walking or even running?" asked Tiger.

"Could we do one with us sitting on Duke and Duchess?" asked Kitty.

"You're not to hold on with your claws. Got it?"

"That's OK, Duchess. If they sit still and we don't run or walk it'll be OK," Duke said.

"Yeah? How are they going to get up there with out jumping and digging the claws in?"

Duke looked at me. *"You could put them up. Couldn't you?"* I nodded.

More pictures were taken as directed. Then I said I wanted some singles of each of them

"You mean all by ourselves? I mean alone?' Tiger asked.

"Yes, Tiger. You never know when a single picture of you may be needed." Like for Christmas or a birthday present surprise, I thought to my self.

I then had them pose separately for me. I told then to pose as they thought they would look best and I added a few suggestions on my own. Duchess was so into regal and she carried it off very well. The tilt of the head up and a bit away, said very clearly, 'look at me'. Duke was standing tall and with his head turned not looking quite at the camera, very princely. He was an elegant animal. Kitty tipped her head a bit to the side but looking right at the camera her eyes were open and there was a hint of a smile on her face. Tiger sat very tall, almost dignified, then raised one paw, tipped his head, and almost smirked. I had to snap the picture quickly before I laughed and he lost it.

"We're done now can we see what we have so far?" Tiger asked.

"So far?" I raised my eyebrow.

"Well, you know just in case none of these work out."

"Yeah. We may see one that just needs a little adjusting," Kitty said.

* * *

I called them when the pictures were ready on the computer.

"Can we sit near your desk so we can see them up close?" asked Duchess.

Tiger sat next to me in my chair and Kitty on the desk. Duke and Duchess stood next to the chair.

"Oh yes, that's good."

"I don't like that one"

"My head is wrong;"

After almost an hour, they weeded it down to two. *"You pick which of these you like best,"* Duke said.

"No, you must choose. It's your present to me. How you want me to see you," I told them.

Finally they picked one out. I was a little surprised but I loved it. Duchess collar showed very nicely and that mattered a lot to her.

"OK," I said, *"now you need to pick out a frame."*

"They won't let us in the store," Duchess said.

"They're some on the computer."

The search began.

"That's too flashy"

"I don't like the color."

They finally settled on a black frame. As Kitty said, *"We're what's important, not the frame. Do you like it?"* she asked.

"Yes, I do. It's what I would've picked."

15
Thieves

We were loading the car. to go to Linda's to ride my new bike. They had a front basket for Kitty and a back one for Tiger, just in case they wanted or had to ride on the bike rather than on the dogs.

"You guys have your leashes? Come I'll put your harnesses on. Come on Duke, get into your harness." Duke was pacing around the car. "

While I tightened the straps, I asked why he was so restless. *"I don't know. I just feel something isn't right. Does Georgia know we're coming?"*

"No, Linda wanted to surprise her. Why?"

"No real reason. I just feel something isn't right. Duchess feels that too."

I'd gotten a bike because when we went for walks on trails they complained I couldn't keep up. They had to keep doubling back to find me. They heard that Linda had gotten an attachment for her bike so Georgia could run along side her. They suggested a bike for me.

When we got to Linda's, Georgia ran over and sat with Duke, Duchess, Kitty and Tiger. *"Are you OK, Georgia?"* Duke asked.

"Yeah, I'm sure glad you came. I've been really worried. Every time the phone rang I'd race over and sit. It worried Linda. She decided maybe I missed you guys. I heard her finally call you."

"Why were you worried?" Duchess asked.

"We were by Dane's and Bejee's house a couple days ago. You remember. The guys we met at the dog show and in the park? Linda stopped to speak to their owner and Dane told us someone is watching the house."

"Watching? Is it doing something or going somewhere?" Tiger asked.

"No, they mean like spying on it?" Georgia responded.

"Oh, like casing the joint," Tiger said.

They all looked at him and Duchess said, *"It's the TV influence."*

"So you tell her," Georgia said. *"We need to go ride by there now,"*

I'd just gotten the bike set up when Duchess came and told me what Georgia said. I nodded. When Linda got Georgia attached to the leash extender on her bike, I asked if Dane and Bejee lived near and could we ride by.

"Yeah, it's a great house. You have to see it." Linda liked great houses.

I tucked the dog's leashes into their harnesses and the cats in the baskets. The dogs ran free by the bike on each side. Hard to tell they weren't attached unless you looked hard.

"How come you're attached to the bike?" Tiger asked Georgia

"She doesn't trust me to stay by her."

"You earned that," Duchess said. *"Or don't you remember running away?"*

"Give her time and never leave her side, she'll catch on. Then she'll let you walk or ride without the leash attached," Kitty looked up at me and said, *"We'll work on it won't we? You can encourage her to try trusting Georgia."*

"There's the house on the right. Oh look, the dogs are by the road. Lets stop so they can all greet each other," Linda said. "We have to ride up the drive a bit, they have an invisible fence."

They all greeted each other. Tiger and Kitty jumped down from the bike before we entered the driveway. *"What's this invisible fence stuff?"* Tiger asked.

"It's an electrified wire buried in the ground," Dane replied. *"See these collars we have on? If we get too close, it shocks us."*

"Oh. That's terrible," Kitty said shaking her head.

"Does it hurt?" Duchess asked.

"You don't want to do it more then once," Bejee responded. *"When they put the collars on us, they had us go near the fence one time. Wow. We jumped a mile. They wanted us to do it a few more times, but we both dug in. They figured we learned."*

"Don't you guys have one?" Dane asked, looking at me.

"Not yet," I replied.

"What!?" I heard in unison. I smiled.

"OK, where are the watchers?" I asked.

"Over there. See them in that driveway down the street? In that truck," Dane said.

There was a dirty white van parked a few houses down. I could see an arm hanging out the window. *"Isn't anyone home there?"*

"No." Bejee responded, shaking her head. *"They change driveways and then leave and come back. Sometimes in the morning, then in the afternoon, sometimes they park on the road. They never get out,"*

"They did yesterday. Our owners were home but the cars were in the garage. One got out and went through the lot next door and then toward the back of ours. Bejee snuck next door so he could followed them through the bushes," Dane said.

"Didn't you get a shock from the fence wire when you crossed it?" Tiger asked.

"No the batteries are dead. We're careful no one notices. They're kinda forgetful, our owners," Bejee said. Then he looked up at me *"You won't tell will you?"*

"You won't run away from home will you?" I replied.

"No way. It's too good here to leave."

"Oh yeah." echoed Dane.

"Ready to go?" Linda asked.

"Linda, I've been watching that truck. The men in it aren't working, just sitting in it. I don't see any lawn equipment."

She looked. "You're right. Ugly truck, too."

"Why don't we ride around some more and keep an eye on it. You don't suppose they live here, do you?"

"In Marsh Creek? Never! We don't see trucks like that parked in front of someone's drive unless it's workers. Well almost never. Let's go and ride. I hope Dane and Bejee are OK. I'll never forget the dognappers."

"Neither will I." Georgia said. She'd been stolen last summer, escaped, then was found by Duke.

We rode around the neighborhood trying to keep out of sight of the truck. It never moved nor did anyone get out. "That's it," Linda said. "I'm calling my neighbor's son, who's a deputy. I'll tell him what's going on and ask him to come by. Now. After all, his dad's dog was stolen. When they came back to get Georgia the second time, he was here and all the dogs caught the dognappers. I'll tell him I'll just sit right here till he comes."

"Wow, she makes moves when she wants to doesn't she," Tiger said.

"You don't know the half of it," Georgia mumbled.

Linda flipped shut her cell-phone, smiling. "He'll be right here. I said I'd get the license plate number and he rather forcefully told me not to go near that truck," she said nodding and smiling. "I knew that would get his attention."

"Oh boy." Duchess said.

"Yep." echoed Georgia.

"Pretty smart, I'd say," Kitty responded.

I agreed. "Good show Linda. How long will it take?"

"Not long, he was just out on AIA. In fact he was coming to see his dad. Let's park the bikes here and walk a bit closer. Over behind those bushes near where Dane is.

"Yeah. Lets go," Tiger said, hopping out of the basket, Kitty following.

"Hold it. No racing away. Linda will have a fit. Walk close by us."

We crouched down and Dane and Bejee were within easy hearing distance. Duke and Duchess told them what was happening.

Soon a sheriff's car came up behind the truck and stopped. After a few minutes two car doors opened. The deputies got out and started walking to the truck. The driver of the truck opened his door, jumped out and ran toward us and then turned into Dane's yard.

Mistake! Dane sprang at him and the man kicked at him. "Duke, Duchess, get him," I yelled out loud. The Shepherds lunged at him and he fell. They stood over him growling while Bejee was pulled his pant leg. Georgia joined the fray and was barking and snarling.

One of the deputies ran up, saw us and then the dogs on the man. "Call them off," The deputy yelled.

The dogs all immediately backed off and sat. The deputy cuffed the man. His partner had gotten the other man who stayed in the truck.

When he had them both secured in the police car he told us when he'd checked the plates. He found there was an arrest warrant out for both men. They were break-in artists. They were casing the joint, as Tiger put it.

Later at Linda's, after she called Dane and Bejee's owners, she told us they had valuable art in the house. The owners had fired a maid a while back, for stealing. Maybe she was related to one of the men. The dogs were stretched out with the cats while Linda and I talked.

"I sure wish I could call you guys when I need you. I tried to think of you, real strong, for a few days," Georgia said.

"Maybe, Duke," Duchess wondered, *"that's what was making us so restless."*

"You think," Georgia quietly asked, *"we could project a message, if we thought really hard?"*

"I think it would have to be a really important message. Like a call for help or something," Duke answered

"Why don't you and Duchess talk to her on the way home," Tiger suggested *"She's smart, she'll know. Anyhow it can't be any more weird than talking to birds and angels,"* Tiger said.

16
Can You Hear Me?

"Excuse me."

"Hi Tiger," I replied.

"Can I sit with you a while?"

"Sure. Hop up." I closed the book I was reading and made room on my lap.

"What are you reading?"

"I'm reading about telepathy, like you guys asked me to."

"Tele... what?"

"Silent speech."

"Like we talk?"

"Yeah. I'm trying to see what they say about sending silent messages to someone at a distance."

"Great. What did you find out?"

"Nothing."

Tiger eyed the book. It was a big book and very thick. It was called *The Element Encyclopedia of the Psychic World. "Awful big book, to tell you nothing, huh?"*

"I don't think they know much. They talked about animals that can find their way home over great distances if lost. Whales that communicate miles under the ocean, using some kind of sonar, but, they're not sure how any of it's done." I ruffled the fur around his ears. *"What have you been doing?*

"Duchess, Duke and Kitty are out practicing calling to each other."

"And."

"It ain't working for them."

I thought that was a funny way to put it. *"Does it work for you?"*

He looked up at me. *"What do you mean 'work for me'?"*

"I mean can you call someone from a great distance?"

"If I had to."

I was silent for a while thinking about what Tiger was saying. *"Did you tell them, you can do that?"*

"I tried to, but they just blew it off, like I was joking or pulling a leg." He was silent for a minute then said, *"That's why I came up here to be with you."*

"Will you tell me about it?"

"You sure you won't laugh or think I'm just telling a big story?"

I knew Tiger never lied or made up things. It just didn't seem to be in the make-up of any of the animals. I wondered if that's why some people say they like their pets better than people. *"I won't laugh and you never make things up."*

"OK. When I was little, a baby like Kitty when we got her, my mother taught me how to call her if I was in trouble. She would go off into the woods to hunt and I would stay in the house or under the porch. As I got bigger I played in the yard near the house."

"Did the man you lived with feed you?"

"Yes, but mother hunted anyhow for fresher stuff. Like field mice and lizards. Anyhow, she told me to hide under the porch and to yell real loud in my head for her. You know, like we are talking now. Not a meow or that kind of thing, just a loud scream for help. She said she would hear me no matter how far away she was."

"Hum, did you do ever have to do that?"

"Yes. The first time I did it, some terrible big dogs, the man called them coyotes, came into the yard. I hid under the porch, way in back, cringing against the house. I was really scared so I yelled for my mother. I watched them and soon they went away."

"Did your mother come?"

"Oh yeah. After they left. She wasn't really mad but she was very serious."

"Why?" I asked.

"She said she hid in the bushes and watched them till they left. If she had come to me they would have killed her dead. She then showed me that they couldn't have reached me under the porch. It was too close to the ground for them to get under, even if they knew I was there."

"I guess that was why she told you to hide there," I said.

"Yes. But I was little and didn't know that. She sat me down and told me how important it was that I never call unless it was really needed. She said if I called for any little thing she would never know when it was important. I could be

ignored and have a bad thing happen."

"We call that crying wolf, " I said.

"Crying what?" Tiger asked.

"Wolf." I told him the story about the boy who called for help a lot when he didn't need to. When he finally did need help and called no one believed him. No one came and he was killed.

"Yeah. I think that's what she meant. I never called again unless it was needed."

"Could you call someone if you weren't scared?" I asked.

"I don't think so. Well… maybe. I wouldn't have to be actually scared. I mean physically scared of being hurt…" he paused. *"If I was afraid something bad was going to happen, that I would need help, or someone I care about would, than I think it would work."* He nodded his head, *"Yeah, it would work then."*

"From how far away would it work?" I asked.

"I don't think it matters. What seems to be important is who you call and why. I mean if you just call anyone for anything it would be awful noisy out there."

Suddenly we heard, *"Come. Hurry. Duke needs us,"* Duchess had yelled at us as she raced from the dock. She and Kitty were both there to see if they could hear Duke call them.

Tiger was off like a shot, me, not too far behind. They were down at the fence by the road. Duke was sitting with his foot held up in the air. *"What happened?"* I asked.

"I stepped on something that got way up in my paw. It hurt real bad. I guess

I yelled real loud in my head. I didn't bark," Duke said. *"It still hurts."*

"He tried to hold his paw up, but I couldn't see anything," Kitty said. *"Can you help him? He really hurts."*

"OK Duke, lay down and turn your paw up to me."

I squatted down, took hold of his paw and started to separate the pads. He pulled back. *"You gotta hold still. No pulling back."*

"OK."

He had a spikey Sweet Gum seedpod stuck between his footpads. The ends of the spikes were really sharp. When he stepped down on, it pushed it into the soft tissue between his pads. I gently separated the pads and pulled it out. He pulled his foot away, sat up, licked it and carefully put his foot on he ground.

"Does it still hurt, Duke?" Duchess asked.

"No. She fixed it."

"I'm surprised it hasn't happened before. There are some up by the waterfront too. They won't hurt the surface of your foot pads cause they're too hard but you got one wedged inside."

Looking at me Tiger said, *"Well it's a good thing it happened,"*.

Duchess started towards him and Tiger quickly scooted over by me. *"I mean it proved what I was telling her,"* he said looking up at me. *"Doesn't it?"*

"Yes, it certainly does. You guys can stop practicing being heard. Tiger explained to me how and when it works. Lets go up to the house and he can explain it."

Tiger told them what he had told me and then how we just discovered pain works that way too. They had lots of questions and finally realized they couldn't just call someone because they wanted too.

Duke looked thoughtful, finally saying, *"How can we help Georgia? I mean how can she call us?"*

Duchess replied, *"We'll just have to sit down with her and explain what we learned."*

Kitty added, *"We're going to have to talk about what kind of reasons we will call for help. I mean, so we're all together on this."*.

"I wonder how come Tiger and I didn't hear Duke yell with pain, yet Duchess did?" I asked.

"I think because he's my brother. I'm more sensitive to him," Duchess replied.

"I bet you're right, Duchess." Kitty said.

Tiger nodded his head, *"I think we have a lot more to learn about this distance stuff,"*

We were all nodding right along with him.

17
Coyotes

"There's a wonderful place where we can go and ride. I was there some time ago with the owners. They gave me the combination to the gate locks and said I could go explore."

"Can we all go?" Duke asked.

"Yes, but you will have to be careful and stick close together and not get too far away from me. This place is a bit isolated and may have more creatures than the public parks and woods where we've walked."

"I'll bet it's like where I used to live with that old man," Tiger said. *"It had lots of creatures and my mother told me to just get in a tree and quietly wait till they went by."*

"OK, go get your riding harnesses, collars and leashes while I get my bike ready."

I fastened my bike to the car, put water and snacks in the saddle bags and was bending over to check the harnesses on Duke and Duchess, when Tiger asked, *"How come you have a gun?"*

My shirt had ridden up and he saw it on my belt. I always carried a gun in the woods, on the boat, and in the car. He just hadn't noticed it before. Duke and Duchess knew and had talked to me about it not long after I got them. *"It's to keep the bad guys honest,"* Duchess said, in a smug knowing voice.

"Would you really shoot someone?" Kitty asked.

"Sure she would," Tiger said. *"Why have a gun if we don't use it?"*

"We?" I asked.

"You know, it's what they call the 'universal we' on TV."

I sighed. *"Kitty, Tiger is only partly right. I think of it as..., for emergencies only. In the woods it would be only to scare off big creatures. OK, lets go."*

We pulled in past the gates and around a group of trees. It was a heavily forested area with lots of upland wetlands mixed in with Silva culture (tree farms). I had explained this to them on the way over.

They stayed on the road near me. It was very quiet, the wetland trees were thick, up against the road but soon the birds and other wood creatures all started scampering through the palmettos and up in the trees. As the trees thinned you could see paths between them. Kitty had been walking beside Duke when he said, *"Hop on, Kitty. Lets go explore."* He looked back at me, *"OK?"*

Off they all went with Tiger loping along with them. I wasn't concerned. They'd been in the woods alone before I got them, as well as with me. They knew I would ride along the road watching the birds and just looking. Duke told me they always knew where I was.

"It's so quiet and peaceful, Duke," Kitty said. *"I just love it here."*

"Yeah. Kitty, look at the little rabbits over there under those stumps. See their noses are just barely sticking out."

"Better stick them back in before some one hungrier than us sees them, " Duchess said.

"Boy, I'll say. What a tasty bite they would be," Tiger exclaimed.

"You wouldn't kill and eat one would you, Tiger?" Kitty asked.

"Sure I would if that's all I had to eat. Kitty, that's what animals do. My mother went hunting for food all the time. That old man couldn't always be relied on to feed us."

"Oh," she replied.

"Hey, lets run over there where it's cleared out a bit. See by those short stumpy trees."

"Those are scrub oaks," Duchess said. *"Like the ones near the ocean. She told us their names.*

Kitty hopped off Duke and walked over to the trees and looked up. Just then they heard lots of commotion in the trees off to their left, and two dog-like creatures burst thru the bushes.

"Climb the tree Kitty, Tiger. Hurry," Duke said, as he turned to face the charging animals.

They skidded to a stop about ten feet from Duke and Duchess. They silently looked the dogs over and then peered up into the very low trees behind them at the cats.

It was very still. Then there was more rustling in the bushes and trees and out came three more of the creatures. The first two were a bit bigger then the others but still smaller then Duke and Duchess.

"Looks like you're out numbered," said the lead animal.

"Think so?" replied Duchess. *"But we're a lot bigger."*

"What are they, Tiger?" whispered Kitty

"They're coyotes," he replied.

"Look skinny, sort of like scrubby dogs to me," Kitty said

"OK," said the lead Coyote. *"We'll share."* He paused and looked up at the cats again. *"These are our pups and this is their final hunting day lesson. Tomorrow they have to go off on their own to forage. So lets share with them."*

"No, that won't work. These are our cats and we're not sharing," Duke said forcefully.

"You got that right," echoed Duchess.

The coyotes sat in a semi circle around Duke and Duchess, who were backed up to the tree. The coyotes looked at the cats and the tree and one said, *"Not much of a tree, one good jump and I'd have them."*

"Oh, Oh," Tiger said. *"Time to call the cavalry."*

"She's too far away," Kitty whispered. "*We don't want her to come running in unaware and get attacked."*

"I'll call her in my mind like my mama taught me. I'll tell her it's coyotes and to bring the gun. So OK, here goes."

I looked at my watch it seemed they were gone longer than usual. They were very good about racing back and keeping track of me. I looked ahead on the road and realized they had gone off into the trees. I rode ahead and then felt uneasy. I stopped and looked around it was very quiet. I could almost feel my skin crawling. What in the world?

I barely heard a whispering voice. *"Come on, hurry."*

I thought, hurry. No, you're hearing things.

"Coyotes."

What? Coyotes? Lord, girl you're losing it.

"Coyotes. Bring gun. Hurry."

Oh my God, that's Tiger calling me. *"Where are you, Tiger?"*

"She heard me Kitty. She heard me," said Tiger.

"Through trees on your left. Bring gun."

"Tell Duke to bark once," I said.

"Duke, bark one time," Tiger said.

Duke let out a loud bark. The coyotes started to stand up. The biggest one said, *"Enough of this stuff. You move aside or we'll tear you apart and get the cats as well."*

"That's stupid," Duchess said. *"What will your kids learn from this."*

"They gotta learn to fight some day. This is as good a time as any."

There was a rustling in the bushes and movement. *"Look Duke,"* yelled Tiger.

They all looked as a human came through the trees. She had a gun and aimed it in the air and pulled the trigger.

The coyotes jumped and raced off. The cats and dogs hunkered down and held their paws over their ears.

Kitty, finally looking down from the tree said, *"That was really loud."*

"Yeah" Tiger said. *"I'd run too and not come back."* Looking at me he said, *"You heard me didn't you? You heard me."*

"Yes, Tiger I did. At first I wasn't sure what I was feeling or hearing, but then you came through loud and clear. You OK Duke? Duchess?"

"Yes," Duchess replied. *"But it's a good thing you came I think they may have been stupid enough to go for us."*

"I agree," Duke said. *"They may have been able to grab Kitty or Tiger while keeping us busy, but I'm not sure all of them would have been alive to tell about it."*

"That's why we have a gun," Tiger said, nodding his head.

18
Dog Overboard

"Hurry Hurry, there's a dog in the water," yelled Duchess.

I looked out the window. Duchess was on the dock looking in the water, surrounded by Tiger and Kitty. I couldn't see Duke, so I hurried out to the dock. The tide was almost out and Duke was standing in very shallow water helping a medium sized black dog to shore.

"Can you help me? Duke asked. *"He seems very tired and will need help to get up the stairs,"*

"Wait till I get my mud shoes on." I turned to Duchess. *"Will you get me some towels from the downstairs bathroom. I'll use the shampoo that's out here."*

"Is he dirty?" asked Tiger.

"Oh yeah. That's half-mud and half-sand he's wading through and he'll be salty as well."

For the life of me, I really couldn't tell what kind of dog he was. Duke and I got him up the steps. Duke sort of pushing and me pulling and half carrying. The dog was doing his best to help. I moved him off the bulkhead over to the dock. I had been patting him and talking to him all the way.

I told the dog I was going to wash the salt water and dirt off of him and all he had to do was rest quietly. I started rinsing him with the hose and he put his mouth over and drank from the hose, slurping and lapping

"Thirsty I guess," Tiger said. *"How come he's not talking to us?"*

"I think he's shy and scared," Kitty said. *"Give him a few minutes. He'll talk."*

I washed and rinsed him twice and then started to fluff and dry him with the towels. *"Oh look he has white paws and some white on his chest. He's a pretty boy, soft fur and fuzzy,"* Kitty rubbed up against him and gave him a purr.

When he was dry and drank more water, he sat and looked at each of us. Bowed his head and said, *"Thank you."*

"What were you doing in the water?" Tiger asked.

"I was over in those woods. I swam there when I fell off the boat." He nodded across the canal. *"I've been there for two days. I couldn't find my way out to where people might be."* He paused. *"I was thirsty and scared. Especially at night."*

"How old are you?" Kitty asked.

"Old?"

"See," Tiger inserted with some authority. *"I didn't know what that meant either. How many warm times and how many cold times do you remember?"*

"Oh. One short, really cold with snow time and then this warm one," he responded.

"May I see your teeth?" I asked.

He looked at me and whispered, *"Yes."*

We all looked. *"Half grown like Tiger when we got him,"* Duchess said. *"Goin' to be a fairly big one, too."*

"Why were you on a boat?" Duke asked.

"The people I belong to were sending me home on the boat. They had to fly off somewhere."

"Where is home?" I asked.

He looked around and nodded to the north *"That way I think. The place where it was really cold and had snow"* He paused. *"I don't remember much. I was very little when we went to where it was warm."*

"Do you know what he is talking about?" Duchess asked.

"I'm not sure, but it seems he was born up north, New York, Massachusetts or somewhere like that. Then when the snow came they went to a second home somewhere in south Florida for the winter."

"Let's take him to the vet's to see if he has a chip," said Duke. *"Like we did with Georgia."*

"Yeah, it's still early. Lets take him now," Duchess said.

After I called the veterinarian, I put everyone in the car and off we went.

Duke said, *"I found him. Can I come in with you?"* The rest nodded that that would be OK.

The Veterinarian, who's my next-door neighbor, saw us right away. He checked but there was no chip. I told him how Duke found the dog and what probably happened. He agreed. He said I should keep him for a while. Surely someone would try to find him. After all he was a very expensive animal.

I heard, *"Expensive?"* from Duke.

The vet continued, "He's a half grown Portuguese Water Dog. You can pay well over two thousand dollars for one and even more depending on the line,".

"Wow!" Duke said. *"Bet he doesn't stay lost long."*

"I'll bet the person driving the boat is having a fit. Especially if he's an employee of the owner. If it happened at night he may not realize it till he's miles away," the Vet said. "I'll listen for rumors. You don't want to advertise. A million people will call you. Pay attention to the news and papers. I'll bet you'll hear."

We got in the car. Duke couldn't wait to tell the others what he learned. *"How did you fall off the boat?"* Tiger asked.

"It was a very big boat and there were two drivers. They didn't stop except for gas. One drove, one slept. I was bored so I went out of the cabin and lay down on the cushions on the back to see the stars. I fell asleep when another big boat went by. I was tossed around and then I was in the water. When I surfaced everyone was gone." He paused. *"So I just swam to shore."*

"Didn't you see lights from the houses?" asked Duchess.

"Yes, but when I swam over, there were only high walls so I had to go back. I went into the woods." He put his head down and dozed off.

The animals had a conference and decided we should keep him, *'if she would let us'*. They put it to me with all kinds of reasons why it was a good thing. I listened and said, *"We'll try and find his owners, and until we do he can stay."*

"Oh," Duke said, *"his name is something really long, like three names, but they call him Pip for short."*

"He looks familiar I'm sure I saw a dog like this on television," Tiger said, eyeing Pip.

"On your animal comic shows?" asked Duke.

Tiger shook his head *"I think it was on one of those programs she watches. You know the really early morning one or the night time ones."* He looked at me.

"I watch Good Morning America and the evening news."

"OK, we'll watch those and maybe I'll see him again," Tiger said.

"That'll be hard if he's here," Duchess said sarcastically. *"I think that's just*

an excuse for you to watch more TV."

"OK, that's enough. We'll all watch the news shows and see."

Four days later I was in the kitchen getting more coffee. I'd been watching the GMA show. Tiger and Pip watched all the shows with me.*" Hurry, they are going to talk about dog safety, Hurry,"* Tiger yelled.

The two anchors both had dogs and they were talking about how to keep them safe, by having them chipped in case they got lost. The woman was saying a friend of hers lost their dog. He fell off a boat in the intracoastal waterway somewhere in North Florida. He wasn't chipped. They showed a picture of the woman with the dog.

"That's my mom, that's my mom," Pip cried, pushing his nose up against the screen. *"They said if anyone sees him will they call the GMA station. A number was on the screen."*

"Hurry, call," both Tiger and Pip said.

I called the number, got an answering service. I left the message saying where I found a dog like the one they showed, and left my name and number.

"Now what?" they both asked.

"Now we wait."

"What if they don't call?" asked Tiger. Duchess, Duke and Kitty had joined us.

Duchess said, rather cynically, *"At two thousand dollars a head, they'll call."*

"How long should we wait till you call again?" Tiger asked. He had taken this on as his project.

"Till noon anyhow," I replied.

"Soooo long," Pip said sorrowfully.

I mentally agreed with Duchess. The dog cost a lot and someone had clout to get this on TV. They'll call.

At 11:30 the call came. A woman said she worked for GMA and was sorry it took so long to return my call but they had a lot of responses to the program. She asked where I found the dog. To save time I told her where, how, when, the breed, what the dog looked like and about how old he was. She said, "Oh, Oh my. I'll have someone call you right back."

"Well?" asked Tiger.

"They'll call back. I think I surprised them."

I had taken a picture of Pip alone and with Duke and Tiger. The latter, with much suggestion from them. The picture was on the computer. I figured if the right person called I would send it and save everyone a lot of time.

The next call came from a different woman. I repeated what I had said and added that he was taken to the vet and did not have a chip. There was someone else in the background talking. I asked if they had a computer nearby, and I would send two pictures. I did.

"Well what's happening?" asked Tiger.

"It's sending."

I heard the woman shout, "That's my baby, that's my baby. Oh, Bob, that's Pip."

"Well, Pip, I guess that is your mom."

The man asked what airport was nearest to me as they would be flying their plane down.

Six hours later a car came up the drive. We were all lined up by the carport waiting.

The door opened and a woman jumped out. "Pip, Pip." she cried as she ran. He raced to her and there were lots of hugs and kisses. Finally Pip turned back to us and said, *"Will you introduce them to my Mom?"*

I stepped forward after introducing myself I turned to the animals. "That is Duke, he is the one who pulled Pip out of the water." Dukes stepped forward and lifted his paw. Pip's mom looked surprised but gracefully shook Duke's paw, as did her husband. This process was repeated with each of them. The cats were obviously a real surprise to Pip's owners.

Then Pip went toward Tiger, *"Tiger, come stand by me please."*

Then turning to me he said, *"Will you tell them how Tiger helped to find my mom?"*

"I'll try Pip."

I think Pip's trying to tell you that Tiger is the one who sat in front of the TV with him, watching the news every morning. When they saw the picture of Pip, they came to get me to look."

"Oh!," she said.

I quickly added, "You know animals are real smart they just can't verbalized to us, so you have to develop signals to communicate with them and they with you. They understand you better then you understand them.

"Oh boy," said Duchess, *" I hope you didn't blow it."*

" I want to come visit, how do I tell her?" asked Pip. He lay down by Tiger's side, and actors that they were, Tiger rubbed on him both looking sorrowful.

"What are they doing?" Pip's mom asked.

"I think he is telling Tiger and us that he will miss him."

"Really," she said. "If it's OK with you, Pip can come visit."

Pip leaped up, danced about then he and Tiger raced around the yard.

"Anytime," I said.

"OK, Pipallion Roman Stophof. It's time to go home."

"No wonder he prefers Pip," Duchess said.

19
Baby In The Water

"Can we take the boat into those little channels toward the Guana Preserve?" Duchess asked. *"You said you used to fish there at the high falling tide."*

Tiger added, *"yeah, that would be a new adventure,"*

"There's no where you can get out of the boat, like we do at that island.," I said. *"It's thick and marshy near the shore."*

"That'll be OK. We'll be exploring strange new water places. Huh Duchess?" Tiger was looking for support.

"Yeah, we can see it from the Intracoastal but we haven't been way back in there. Looks mysterious." replied Duchess.

"What's mysterious?" Tiger asked.

"Sort of strange and new," I replied. *"I'll pack some snacks for us and some water, we'll have about 5 hours of high tide."*

* * *

We were on a rising tide and I had slowed down to enter the first of the cuts (small inlets) when Duke yelled. *"Look over there. Two heads in the water on the right, see um?"*

I looked ahead to my right and there was a deer in the water and another smaller head. near it. "Yes, *I think it's a baby, a fawn.. It's struggling a lot."*

I was barely moving thru the water to the deer and indeed the fawn was almost submerged and sputtering. I stopped next to it, reached down and hauled it into the boat. *"Is it OK?"* asked Kitty.

It could hardly stand and was shaking a lot. I wrapped it in a towel and held it for a minute. *"Look,"* cried Duke. *"The mother deer is still near us."*

She was out of reach, about 15 feet from the floating boat. We were nearing the shore. She swam closer to the shore where she could touch bottom. She stood and watched. Then we heard gun shots from the across the marsh and toward the trees to the west. The mother deer swung her heard around and looked toward the the sound.

"Kitty, can you talk to her?" Duchess asked.

Kitty made some noises and the doe did as well, this went back and forth a bit. *"Well what did she say?"* Our always patient Tiger asked.

"She said there were men with guns shooting at things over in those woods. They had to get to the water to get away. Her baby is very young and got tired and couldn't swim any further.".

They talked some more. *"She is really worried. She wants to know what you will do with her baby?"*

"Good question," Tiger said.

"Well we can't put it back into the water. It won't be able to go in either direction safely," Duchess said looking all around.

"Kitty, have her go stand on the shore and I'll bring the boat up and beach the front. Then we can talk and decide. You can ask her more about the men. Where they came from and that kind of stuff."

I beached the boat at an angle to the beach with the out drive up to keep the propeller off the bottom, while we continued to talk.

"Why do you care about the men?" Tiger asked me.

"Cause there's no hunting over there and it would be hard for them to get in, shoot guns and not get caught." I replied.

"You think they got there by boat?" Duke asked.

"Hey," Duchess said *"poaching by boat. If someone comes looking for where the guns shots came from, they just take off in the boat instead of getting caught on the roads."*

"You're right, Duchess. Who would ever suspect a boat . No one would look for dead deer in a boat,." Duke said. *"You know that place we ran away from, we heard the men who worked there talking all the time about that being a great way to poach and dodge some kind of law men."*

I thought about what they said. We needed to get these deer out of here, before someone came looking for them in a boat.

"Kitty, ask the Doe if she can get into the boat. Then we can take her somewhere safe."

We were sitting sideways to the shore. The Doe walked up to the boat looked carefully at her baby cuddling in a towel in the boat. She and Kitty exchanged some words.

Kitty then said, *"If we all move to the front she thinks she'll have enough room to leap into the back section."*

"Kitty, tell her I'll stand close by in case she overshoots or stumbles. and catch her," I said, as I pulled the baby further toward the front to give her landing room.

She wasn't large, one leap, a bit of a stumble and she was in. I told her to lie down and the dogs to sit up front with Tiger. Kitty stayed by the deer. I told everyone I was going to go into the cut where we had been heading. and we'd see if a suitable landing place showed up.

"Wow this is nice," Tiger said to me.*" It really still. Twisting and turning around back here. I can see why you like it."*

"Yeah," Duchess added, *"but I'd hate to fall in and have to figure out how to get out."*

Duke looking around added, *"There's nowhere to stand except on those oyster mounds. That wouldn't feel so good on the feet."*

The doe was peering over the side of the boat looking all around. There was the shore of the Quana but next to it was a deep stretch of marsh and water. We looked further but it wasn't much better. The doe made some sounds and Kitty said, *"She thinks she maybe able to force her way through the marsh and water but her baby couldn't."*

"What are we going to do?" asked Tiger.

Duchess looking all around said, *"Let's take them down to the bluff where we helped the cormorant. We were able to beach the boat there,"*.

"Yeah Kitty ,you tell her about it," Duke said.

"Kitty, tell her to sit up a bit more so she can watch where we are going and

I'll explain as we go."

"Just make sure she ducks down when we tell her to." Tiger said. *"Duke and Duchess and I will be lookouts for gun guys as we go."* He paused. *"What an adventure."* Duke and Duchess both nodded their heads.

After we left the cut and headed south. I gave a running commentary to Kitty who relayed it to the doe. She watched the far away shore as we went south it was still all deep marshland.

"There's our island!" exclaimed Tiger. *"Won't be long now."* He paused. *"Oh Oh! Look there's a boat creeping out of that cut over there. Get her down and cover her up with a towel."*

Two guys in a boat, I didn't see any fishing poles. I looked away and tried to keep on going. But they cut over toward us waving and yelling. *"Duchess, get in the back and when I say so you and Duke bark and snarl. Kitty, tell the doe not flinch or move and not to be afraid. Tell her Duke and Duchess are going to make a lot of loud noises to scare the men away. Then, you and Tiger get here by me*

As the boat drew neared I slowed a little, "Hey" they yelled. "You ever see any deer in the water north of here."

"Deer in the water? No, I don't think so. Years ago I saw one."

"Those are nice dogs mind if I come closer?"

"They aren't very friendly," I responded. But the men kept edging closer.

"Show them how friendly you are Duchess, Duke."

They started with a snarl, all teeth showing, that progressed to full scale barking, then lunging against the side of the boat, as if they were going to jump. *"Good show,"* Tiger said.

"I told the doe, again not to move, that Duchess was scaring them away," Kitty said. *"Scares me too. I wouldn't want to be on the receiving end of them."*

The men stopped their boat and veered away. "What the hell lady, damn, they're vicious."

I could barely hear them over the noise of the dogs. "Just doing their guard work," I yelled. I saw one of them reach into the bottom of the boat. *"Duke, Duchess stop, now."*

I saw what looked like the stock of a gun by the man's knees. "Sorry for the noise, they're very protective. Bye, good fishing." I pushed the throttle up and we left them looking back at us, then they turned north.

"OK, we safe. They're going north where we just came from. Ha! Too late bad

guys, we got away." Tiger was nodding his head so was Duke, Duchess, Kitty and even me.

There were no boats at the shell bluff so I slowly beached the boat and edged it a bit sideways. I threw a rope on shore and asked Duke to get out and hold it so we wouldn't float off. There was a very quick drop off and a good current on this rising tide. The doe leaped out. I got out, lifted the baby out of the boat. She was a handful but she didn't kick. I heard her mother talking to her.

I told Kitty to tell her which direction to go in. Duke asked, *"Could we follower her for just a little bit to be sure she will be OK?"*

Since they had their harnesses on and Kitty could ride and Tiger too if he wanted I said, *"OK but if you see or hear people, race back here fast."* I didn't know if dogs were welcome in the park.

It wasn't long before they came racing back. *"We heard people in the distance. The doe and baby raced into the woods and we came back,"* Duke said. *"Wow, what an adventurous day. We saved another baby.'*

"Yeah," said Tiger. *"Now can we have snacks?"*

20
Snake Bite

I had just put the loppers in the cart in preparation to trim the palmettos on the north side of the yard by the fence. Palmettos were scattered along the fence line in several places but they were really heavy on the north side. An assortment of wide leafed plants, ferns and azaleas were planted among them. It was an area I let grow wild except for once a year when I trimmed some of it especially the palmettos.

"Can we help you?" asked Duchess.

I explained what I was going to do. They were very good at dragging palm fronds about. I wasn't sure how they would do with the big palmetto fonds. I said, *"Sure, do what you can. They're kind of long and awkward to handle, but I'll take all the help I can get."*

I started on those nearer the grass and was working my way back to the fence. Duke and Duchess were actually doing a good job of dragging them and lining them up by the cart. Kitty was supervising the dragging and Tiger was near me and watching the under growth that I was moving though.

I heard a shout. *"Duke!"* The next thing I knew I was flat on my butt. Tiger had seen the snake coil and move to strike. Duke had leaped between the snake and me, knocking me away.

Duke cried out and twisted away. It *bit Duke,"* Tiger yelled.

Duchess was right by my side and Duke was limping as he came near. *"Is it gone?"* asked Duchess.

Tiger was looking carefully at where the snake had been. *"I don't see it. I think so."*

"You guys get back," It looked to me as if Duchess was going after it.

I examined Duke. He had been bitten high on his right leg. He was holding his leg up. *"Lie down Duke, I'll back the car up here and help lift you into the back."*

"Why?" Kitty asked. Her eyes big and frightened.

"I have to get him to the doctor immediately. Did you see what kind of snake, Tiger?"

"I heard a noise first. I think it was a rattle snake?"

"Is that bad?" Kitty whispered.

"It'll kill you.," Tiger declared.

"Not always, Tiger. Don't scare her. But it is serious. That's why we have to hurry."

I grabbed the car keys, my wallet and backed the car up by Duke. With his help, I got him in the back of the wagon. Everyone else piled in next to him. I had called the Vet and told them what happened, that we were on our way. They said they would meet us and carry Duke inside.

"Does it hurt a lot, Duke?" asked Kitty.

"Yeah, Kitty. It hurts all around it too," he replied in a soft voice.

"Duke, you lie still. Don't be moving around at all," I said.

"We'll watch him," said Duchess.

* * *

We parked by the side door. The Vet and a helper came out and lifted Duke out of the car. We trooped right in after them. Duchess, Tiger and Kitty scrunched up tight behind me, and lay against the wall, like no one would see them there. The Vet and his assistant put Duke on the examining table and had an IV hanging on a hook. Duke was panting and trembling from the pain. The Vet looked at me and said, "I don't want to give him an anesthetic. Can you get him to hold real still while I start an IV? I have to give him antibiotics and the antivenin medicine.

"Yes. I can keep him still." I held Duke and said out loud, "Be still Duke, don't move at all."

He opened an eye and said, *"OK. Will it hurt?"*

"No more than your old shots. They're going to give you some shots and then put a needle in your leg with some tubing attached to it to give you fluids and more medicine. Whatever you do, don't move or pull it out. I'll be right here."

"We're here too, Duke. We'll watch," Duchess said.

They shaved some of his fur and had the IV in and taped down in a flash. Then they gave him a couple of shots and that was it. "You're a good dog, Duke," said the Vet, petting him on the head. "You didn't move at all." Turning to me he said, "He certainly is a great dog. Does what ever you tell him, doesn't he?"

"Yes, they're all great animals and very obedient."

"I'm going to put him in one of the big cages in back. We'll keep him over night. If he makes it through the night he maybe able to go home in the morning."

"What does he mean, 'if he makes it through the night?'" Duchess asked sharply, suddenly standing up.

I turned and put my hand on her head. "It's OK, Duchess." I then said. *"It's just something they say about really sick dogs and people."*

"I don't think so," Tiger said. *"How bad is Duke?"*

The Vet turned when he saw Duchess and then the other two. "Oh my, I didn't see them.

They were so still. OK, let's carry him back." They picked him up and we all followed. I'm sure the Vet didn't think for a minute they would follow. Once Duke was in the cage he turned and there they were at his side. "Time for you guys to go home."

"I'm not leaving," Duchess said. She planted herself against the cage and didn't budge.

"Come on Duchess, you have to go now." The Vet said.

Duchess looked at me and said, *"I'm not leaving Duke here alone. He's sick and he has never been alone before."*

I stooped down next to her and put my arm around her and said to the Vet, "They have never been separated before, and Duke would be scared if he woke and no one was near."

"I can appreciate that, but she can't sit in the aisle all night and the cage isn't big enough for two big dogs."

"I'm not leaving him alone," Duchess repeated.

"Duchess, I can stay with him, if it's OK with you?" Kitty quietly murmured.

I looked at her and said to the Vet, "How about if we left Kitty with him. She's very small, won't cause trouble and she has been with him since shortly after she was born."

The Vet looked at Kitty and then at Duchess, "Well, we don't normally do such things, but these animals are very bonded with each other. I can see where leaving one alone could be traumatic. Do you think Duchess would be OK with that?"

"Lets see," I said.

As I picked up Kitty I said to Duchess. *"Duchess you can't stay here. He's bending over backwards to help us by letting Kitty stay. That's the best we can do."*

Duchess stood, walked to Kitty and gave her a nudge and a lick. *"You'll watch him real good, yell and call if he needs anything?"*

"Yes Duchess, I'll stay right next to him. If he wakes and get's scared I'll give him licks and rubs. We'll be OK," replied Kitty.

"Do you want me to stay too, Duchess?" asked Tiger. *"I will, you know…"* he paused looking all about with big eyes.

"Thank you, Tiger, but I need you at home with me," Duchess said, knowing that Tiger was scared at the vet's office.

The Vet opened the cage and I carefully placed Kitty in the cage next to Duke. "Kitty," I said out loud, "don't touch the tubes and call loudly if Duke needs anything."

"You know," said the Vet "I think she really understands you."

"I talk to them a lot, give a lot of simple directions. They're also very clever and smart."

As we turned to leave, Kitty called out, *"Duchess, I'll ask the angel to come sit with us."*

On the drive home Duchess asked again if I thought Duke would be OK. I told her the Vet said he really thought so. Especially since we got him to the Vet so quickly. He said Duke was a strong dog and the bite was in his upper leg but away from his chest. That seemed to quiet her.

The afternoon went quickly and late in the day the phone rang it was the Vet telling us Duke was doing well and so was Kitty. He said he would go in early and call and tell us if Duke was good enough to go home in the morning. Duchess seemed a little better hearing that. But she and Tiger stayed right by my side all day.

At bed time Duchess quietly asked *"Could Tiger and I sleep upstairs by you?"*

"Yes, you both can sleep up here. We need each others company."

I could certainly appreciate what people with children went through. Well, they are my children, I guess.

* * *

Early the next morning the Vet called and told me Duke was fine, we could come get he and Kitty at ten o'clock. Thereafter, Duchess and Tiger paced the floor repeatedly asking *"Is it time yet, is it time?"*

They brought Duke and Kitty into the examination room where we waited. Immediately there was much nudging, smelling and licking. "You know," the Vet said, "we had a tech stay last night because we had a three sick animals. He said once or twice when he went in to check on the animals he saw a glow, like a light, shining near Duke's cage, but when he got nearer it was gone. I know he doesn't drink or take pills he's sober man." He paused and looked at me.

I shrugged, "Maybe a reflection off the metal of the cage when he opened the door. Is Duke able to be up and about at home."

"Yes, he'll be a bit sore for a day or so, but he's just fine. I'll have the drugs for vaccinations for the others in a day or so I'll call you."

On the way home Tiger asked, *"Kitty what's with the light? "*

"The angel came and she glowed whenever she touched Duke."

"Did you see her, Duke?"

"No Tiger, I just felt warm and my leg stopped hurting, then I went back to sleep."

"Well I'm glad we are out of there," Tiger said.

"Not so fast Tiger. You all have to go back in a day or so and get a vaccine shot against snake bite," I said.

"No! No shots. I did that already. No more." Tiger wailed.

"Then you can't go outside any more.," I said.

"Yeah, you'll just be house cat, probably grow into a big fat one," Duchess said as she nudged him. *" Don't worry Tiger, I'll be with you."*

"We all will, Tiger," Kitty said.

21
The Painter

"I'm having the house painted this week," I said.

"What's painted?" asked Tiger.

I waited for a sarcastic remark from Duchess, but they all just looked at me. I'm used to them being so smart I forgot this might be a new word to them. *"OK. Painted is when you put a color or substance on a surface."*

"What's a substance?" asked Kitty.

"Yeah, and what is a surface?" echoed Tiger.

Oh boy, I thought. *"Let's see, a substance is a liquid that comes in a can, a liquid like water."* I looked at them, no lights went on in their eyes. I got up and pulled a can out of the cupboard. *"This is a can."* I then turned on the water. "T*his is a liquid."* Then I walked over to the wall, patted it and rubbed my hand on it, *"This is a surface."*

Duchess looked at me very carefully, *"So-o-o painting is putting water in a can and then on the wall."*

I tipped my head to the side and eyed her. No, I thought, she is not being a smarty-pants. They don't get it and I'm doing a bad job of explaining. *"Come with me. I'll show you."*

I went to the large cupboards on the carport where I kept paint. I opened a can and took a stick and stirred the contents and held up a dripping stick and said *"This is paint "* Before they could interrupt I took a brush out and said *"This is dipped in this can and then brushed on the surface of this wall."* I patted the side of the building. *"The process is called painting."*

There was a general nodding of heads, then Kitty looked at me very carefully and asked, *"You're not mad at us are you?"*

"You don't think we're stupid, do you?" asked Tiger.

I felt ashamed and realized not only had I been impatient with them, but I had really talked down to them. I apologized and said, *"I'm sorry for talking down to you like that. I was so used to you all knowing everything I just forgot that there are some things you don't know."*

"That's OK," said Duke. *"We just don't want you to be mad at us."*

Then I felt worse.

"Well moving on," Duchess said. *"what color are you going to paint it?"*

"What is color?" asked Tiger. I just stared at him. *"OK, OK, I'm just kidding, but can we talk about color sometime?"*

"I like it just as it is, white. My home is white. Please don't change it, don't make it different. It just wouldn't be the same, like mine, I mean," Kitty said very shyly and quietly.

Duke walked over and rubbed on her. *"It's OK Kitty. I agree it wouldn't be the same, not like home. I mean home like it is now. Please don't change it."*

I was astonished, and realized I was in full agreement. *"I have no intention of changing the color, this white house is home to me, too. But it needs to be repainted. It's looking dull and I want it to look bright and shiny again."*

"I guess red with some sparkles won't go, huh?" Duchess said looking shyly at me out of the corner of her eye.

I could swear she was smiling. *"Maybe we can have the frame of the door bell painted red, Duchess. I'll think on that. What I really wanted to talk to you about is how you behave with the painters."*

"Why, we'll help them just like we did with the dock man Tom. We were a lot of help, and he liked us," Tiger said with a mighty nod, the others all said, *"Yeah."*

"They will be here the day after tomorrow and we will see after I talk with them. It's not the same, you know," I felt like I was talking to the wall.

* * *

Harry, the painter, came with his helper, Shawn. I introduced the pets and asked if they would be a bother to him.

"Generally no, we like dogs and cats, but today they will have to stay far away or inside. We'll be spraying the house with heavy duty Clorox and then power washing it. It'll burn their eyes and maybe their feet. I'll wash it off the patio and carport when I'm done. We'll start painting tomorrow."

"The dock should be far enough away," said Tiger walking to the dock.

"Yes, we won't get in the bad stuff. He said he'd wash it away later," Duke stated.

"I think I'll stay inside and watch out the window today," Kitty added.

True to their words they were mostly on the dock or they walked far away from the house, near the fences. Kitty moved from room to room and peeked out the windows at them.

"Wow! Look at that ladder. It sure is long," Tiger said looking up.

"Yeah," Duke agreed *"I never thought about how tall the house is till I saw them way up there."*

"Gonna get a crick in my neck looking up so high," Duchess observed.

* * *

The next day the painters started early and true to form the animals met them as they arrived.

"I'm not sure how we can help?" Duchess said.

"I know," Tiger said, *"we can watch and tell them if they missed a spot."*

"Sure can't do that when they are painting way up high. Have to wait till they get lower," Kitty said, *"And that white over white will make it even harder to see."*

"I think we'll do OK. The painted part will be more shiny," Duke said.

"Oh, look they're splitting up. The one she said was boss, Harry, is going up high . Shawn is going to do the upstairs porch. How about Duke and I watch Harry and you two go up to porch and watch," Duchess said.

They removed all the porch furniture as I looked out thru the glass doors. The cats were sitting in the doorway to the landing.

"I can see good from here," Tiger said.

"Me too," echoed Kitty.

"Oh, oh, he missed a spot. Tell him Tiger." Kitty said.

Tiger walked over and patted Shawn on the leg and meowed. Shawn looked down and said "Hi kitty cat, can't pet you now. My hands have paint on them." He turned back and felt another pat. "Later I said." Another pat followed. He looked and the second cat was there and had patted him too.

"Tiger, now pat the spot he missed." Kitty said.

He watched the bigger cat, called Tiger, pat the frame of the glass door frames he had just painted. "Watch it, that's wet paint." The cat held his paw up to him and there was no paint on it. He looked at the paw and then at the door and sure enough there was no paint. He could see he missed that section. He took his brush and painted the area he missed. The cats moved back to the door and sat.

I heard a knock on the glass doors. I walked out to the porch and Shawn told me what the cats had done. I looked at Tiger and Kitty.

"See we're helping," Tiger explained.

"Shawn if they are bothering you I can put them inside."

"No, they're no bother. I did miss a spot. I just wanted to sort of ask are they really smart? I mean, well are they going to watch all the time."

"Not if they're bothering you," I replied.

"No, no, bother, save me time, it's well, it's just kinda strange." He looked at the cats and grinned. "OK, kitties, lets get back to work."

I looked at Tiger and Kitty, *"You two be careful. You're walking a very fine line here."*

"Walking a what?" asked Tiger.

"I think she means we are almost in trouble. We'll be careful, we promise," Kitty said.

I watched them off and on, then I heard a bark in front, and Duke raced around the house and yelled, *"Come help before he falls."*

"Shawn, come with me. I think Harry maybe in trouble." We raced down the steps and found Harry and the ladder tipped a bit side ways. He was propped against the big windowsill of the attic, holding the ladder from falling further over.

"The leg of the ladder slipped into a hole I hadn't seen. I can't get it straight," Harry said.

Shawn grabbed the ladder so it wouldn't slip any more, "Will you get a stone or brick to slip under the leg while I hold it?"

"There's one by the patio," Duchess said.

We both ran and got it and while Shawn held the ladder straight up I pushed the brick under the ladder leg. Harry came down the ladder. "Time for a break, I think," he said and patted and rubbed both dogs. "You two saved my life. Thank you." He looked at me. "That one, Duke barked right away as it began to slip gave me just an instant alert. Just enough to grab the sill and hold it, but I couldn't have held on long. I'm sure glad they got you to come so fast."

"Yeah, Duke is a hero, saved his life, he did," Duchess pronounced.

"Yeah, you wait till I tell you what these smart cats are doing to help me," Shawn said as they walked over to their truck to get a soda.

"Well, are we doing OK?" Tiger asked, after he told Duke and Duchess what they had been doing, while we walked around to the front of the house.

"Tell me again how it works?" asked Duke

"We meow and pat his leg and then point or touch the spot he missed," he said while eyeing me.

"Un-huh," Duchess said and looked at me. *"Do you think he's suspicious?"*

"I don't know what to think. He says they're just smart. I think contemplating more would be a step too far."

"Contem… what?" asked Tiger.

"I think she means no," Duchess said.

* * *

The painting went on. The animals continued to watch and help. Tiger complained there wasn't much to do. He said, *"They sure are good. I think maybe they're being extra careful cause we're watching."*

I smiled, *"That's good to know, but I've used Harry before and he's very good even when not watched."*

The next afternoon I walked out to see how it was going. Harry was washing a bit of paint off Duchess' back and Duke's tail. He laughed when he saw me. "They got a little close and Duke's tail brushed the post."

"Let me do that," I said.

"No, they're my helpers. It's no problem. The cats stay close to Shawn and they watch their tails better. You know they're great animals and good company. I'll miss them."

"Tell him we'll miss him and Shawn, too," Kitty said the others nodded.

"I'm sure they'll miss you too."

"We will be able to finish tomorrow. I'll paint the cupboards in the kitchen, the doors and windowsills. Shawn will do the windows inside and out. Then we'll do the carport floor. Kitties and dogs will have to stay in for 3 or 4 hours after we are done."

The next day the dogs watched Shawn on the ladder outside as he did all windows. Duke and Duchess followed every move, as Duke put it, *"In case the ladder slips."*

Kitty and Tiger followed every brush move of Harry's. Every so often I heard a meow and then a "Thank you" from Harry. Sometimes I watched Harry and pointed out a spot or two. I didn't get a 'thank you.'

"We would have picked that up," Tiger would say.

I returned to my office.

As the men were getting ready to do the carport, Harry looked down at the animals before he shut the door to the car port. "I want to thank all four of you. You do good work. I'll leave so extra carport paint in case we miss a spot. She", he said looking at me "can touch up what we miss tomorrow."

"Tell him we will all look it over and tell you where the spots are," Duchess said to me.

"I can't wait." I replied with raised eyebrows.

22
Thunder, Lightning, and Trees

I sat reading in the big chair by the windows over looking the waterway. Thunder rumble in the distance. The bright sunny day quickly darkened. Large gray clouds were racing in from the west. They were moving fast and were layered, rolling over and under each other. I loved theses fast moving summer storms. The wind was picking up and the tree limbs were starting to dance and the thunder getting louder.

"Hide, here it comes."

Tiger skidded across the kitchen floor to the door. *"Did you hear me? You gotta hide it's coming."* Duchess, Duke and Kitty were right behind him.

"What's coming Tiger? Is someone outside chasing you?" Just then the thunder roared and I saw lightning far off in the dark clouds.

"That's what's coming, see," he nodded towards the window, *"it's really loud. I'm getting under the bed. Come on Kitty! Duke and Duchess will go to their place in the bathroom closet."* Off they went.

I forgot how scared they were of thunder. When they heard it they'd go off quietly and hide till it's over. I was surprised, one day, when I found Duke and Duchess in the bathroom closet. It's a small dark room, I just smiled and didn't say anything. Actually I acted like I hadn't seen them.

I knew they'd be embarrassed for me to see them cringing in a corner. Not exactly a German Shepherd image. I continued to read and watch the storm.

"Tiger, I'm really afraid for her sitting there in that window."

Tiger looked out from under the bed, *"Kitty, she sits there a lot, storms don't seem to bother her."*

"But, I think this one is different. It feels different and that big tall pine tree is right near that window."

Boom! Crash! The sound of palms founds hitting the roof sent them scrunching even closer together under the bed.

"See Tiger, it's a bad one. You go tell Duchess to tell her to move."

"Me? I don't think so. What if it caught me on the way to the bathroom? "

"But there are no windows between here and there. Run quick, Tiger"

"Oh boy. Well here goes." He edged out from under the bed and raced to the bathroom. It was right next door, and squeezed into the closet by Duke. *"Duchess, Kitty says this is a really bad storm and it's not safe for her to sit by that window. She wants you to tell her to move."*

"Me? Who elected me to be in charge?"

Duke ducked his head, like maybe no one would notice him. *"You go Tiger, you're already half way there,"* replied Duchess. Then she nudged Duke, *"Or better yet, you go, Duke."*

"No Duchess, you go she pays more attention to you," Tiger said as he skittered off.

Finally Duke said, is a small voice, *"I'll go if you want me to, Duchess, but she does pay more attention to what you say."*

Duchess slowly got up, grumbling to herself, and headed for the office sitting room. As she turned from the hall into the dining room, Duchess saw her coming thru the kitchen door from the office. Suddenly there was a terrible crack of lightning and thunder together. A huge flash of light lite up the whole house and even seemed to shake it. You could smell the ozone in the air.

"Wow!" I said out loud as I jumped. "That was close." I turned when I heard whimpering. Duchess was crouched on the floor shivering. I went to her and put my arms around her. *"Duchess are you OK?"*

"It almost got you. Kitty was right."

"What do you mean Kitty was right?"

"She sent Tiger to us and told us we had to make you move from that window."

I looked back at the window, I had been sitting next to and thought, good thing I did. The hair on my arms was standing on end. I wondered what made me move when I did. I smiled to myself and said a prayer. Thank you Lord. Must have been Kitty's angel.

"Thank you Duchess, you can go back with Duke now."

"You're not going back to the chair, are you?"

"No, I'll read in the living room till the storm is over."

* * *

After the storm passed I found the TV and the phones had been knocked out, I was told all the electronics were burned. Worse, the tree near the window was hit and had a large vertical split, you could see the yellow bark inside. It needed to be removed before it weakened and fell in the next storm. The tree was almost three feet in diameter and nearly eighty feet tall. After the owner of the tree service gave me a price, he and his helpers were discussing what would be involved. He said it would have to be taken down in pieces. He was walking about looking at the tree and the yard considering where they would drop it.

Duchess, Duke. Kitty and Tiger were looking up at the tree and the house.

Duchess said, *"Look at that big black mark on the side of the house."*

"Yeah, on our new paint job," Tiger grumbled.

"Right near where she was sitting," Kitty said with a look in my direction.

"Could have burned her right up. Yes it could." Tiger stated with an emphatic nod of his head. *"We saved her, we did."*

"Do you think she's seen that black mark?" asked Duke.

Duchess snorted, *"Be hard to miss, unless she wanted not to see it."*

"Duchess, that wasn't nice, it was almost mean," Kitty scolded

I had been eavesdropping on the conversation and knew exactly where Duchess was coming from. My stubbornness had almost gotten us both hurt and scared Duchess badly. She was not the most forgiving of creatures.

I walked over to them and rubbed Duchess' ears, *"No Kitty, Duchess wasn't being mean, just honest. I could hear the thunder getting louder, see the lightning coming near and I knew there was a tall tree by the window. To continue sitting there like that was indeed stupid and dangerous."*

"I didn't say you were stupid," exclaimed Duchess.

"No, you're much too polite to say that but…" I laughed.

"Well it's another job for us to help with, better figure what we need to do," Tiger said, as he walked about looking the situation over.

He and Duke walked over and sat by the owner and looked up when he did and listened attentively to the discussion. The owner watched them out of the corner of his eye. He noticed as he moved, so did they. When he looked up or over at a nearby tree, so did they. He reached down and gave them both pats and said to me "Looks like I have two helpers. Good workers are they?"

"Oh yeah, very helpful," I rolled my eyes.

"I *saw you roll your eyes,"* said Duchess. *"We are helpful and we won't get in the way. Trust us."*

"If they get in your way tell me."

"No problem. I had a dog that came with me on every job. Animals are smart, I'll watch em."

"See, I told you." Duchess said and joined Tiger and Duke.

"What will you do, Kitty?" I asked.

"If you're going to stay out here I'll just stay near you. OK?"

"I think that's a good idea. Safer too. I'll get a lawn chair and go sit under those trees by the water. We can see good from there."

* * *

"That's neat," Tiger said. *"See they put the rope around a nearby big tree limb then tie the end to the limb they are cutting. After its cut they lower it to the ground."*

"I see," Duke said. *"They saw the small branches off first,, then lower them slowly and when they get low enough they grab them and move them to a safe landing."*

"In these tight spots then just can't let them drop anywhere. I'm impressed," Duchess said nodding her head.

The owner walked about eying the next limb to cut. His furry helpers moved with him. Finally he said, "OK lets get that big limb. You two get on the lowering rope with Matt this one will take more then two." He reached down and petted the dogs in the absent minded way that people do who are used to having dogs around. "You guys move off a bit now," he said to the animals as he pointed to where he wanted them to move.

"He's a nice man isn't he," Kitty said to me.

"Yes, you can tell he's used to dogs." I looked at Kitty she was watching every move just like the others. All of a sudden her ears twitched and her hair stood up on her back. She yelled, *"Tiger look at the rope."*

Tiger looked up and shouted *"Duke push him."*

With barely a glance at the rope, Duke jumped at the man and shoved him violently away, as the rope broke and the huge limb fell with a loud crash right where he had been standing.

"You OK Boss, you OK?" The man in the tree yelled.

"Yeah, I'll say." He looked closely at Duke, then reached over and gave him a big hug.

"You saved my bacon that time didn't you boy?" Looking at me. "You sure you wouldn't like to sell me this dog? Damn, he saved my life."

Tiger stick his chest out. *"Yep, we all make a great team. You did good Duke, Kitty too."*

"I could never separate them." I said proudly. "They're a team. My team."

23
Airplane

We'd all just come inside from mowing the yard. I mowed and the dogs picked up any loose sticks that were in the way. I was opening a bottle of water when the phone rang. "Hello."

"This is Becca, Pip's mother." I heard a laugh. "I thought that would tell you right away who you were talking too."

"Pip's mother, is self-explanatory. How are you Becca, and how is Pip? He's not lost again, is he?"

"Pip? Is Pip OK?" asked Tiger his ears alert.

I ignored him.

"No, he's not lost, he's fine. He's just acting funny?" she said.

"Acting funny?" I repeated.

"What's he doing? Is he OK?" Tiger asked again.

"Excuse me a minute," I told her.

"Tiger hush. I'll tell you all about it when I'm finished."

He hung his head, *"I'm just worried,"* he mumbled.

"I know, but hush now."

"Yeah Tiger. She can't talk on the phone and listen to you at the same time," Duchess admonished him.

"I'm sorry. Just a bit of an interruption. The dogs rushed over when they heard Pip's name. Now tell me what you mean by acting funny."

"We're at our summer vacation home on Martha's Vineyard. His cousin Bo came two days ago with his family. I thought that was great and would give Pip someone to play with, as they are only a couple houses down the beach. But after the first day Pip came home and seemed restless. He would just look at the picture of your animals and whine. This happened yesterday too."

"Really," I said and repeated what she said. "Is that what you said?"

"Yes, then I asked him what was wrong. You know how one talks to their dogs almost as if we expect them to answer us. Well, he picked up the picture and walked across the room and put it in my lap."

Again I repeated most of what she said and added, "Has he ever done anything like that before?" They all stared at with rapt attention.

'Well sort of. I'll ask him if he wants to eat and sometimes he'll bring me his bowl. He's a very smart dog."

"Yes he is."

"So I thought, maybe he misses Tiger and Duke and all. I jokingly asked him if he missed them. He barked and ran in a circle and then lay down and stared at me. That's why I'm calling. Would you all come and visit?"

"Way up to Martha's Vineyard?"

"I know, but it's not as bad as it sounds. My husband Bob is in Palm Beach and will be flying home in a day or so. He could stop in St. Augustine, like we did before, and you can fly up in our plane with the animals."

"Fly up to Martha's vineyard with the animals? Well I don't know. They've never been on a plane before?"

"Fly?" I heard in a chorus.

"Pip did fine his first time. You could give them a pill."

"Give them a pill."

"I don't think so," said Duchess.

We talked a bit more and I told her I would mull this over for a bit and call her back.

* * *

"Fly?" asked Tiger. *"What is fly?"*

So I explained about flying and airplanes. Then I said, *"You've seen their lights in the sky at night, remember? You also see them on television."*

"Yeah? On television, I thought that was like the funnies, like not real," Tiger responded.

The others just sat there and listened. Finally Duchess said, *"See. All that television hasn't really made him smarter."*

"Come on Duchess, how do you know so much about airplanes?" Tiger asked. She'd hurt his feelings.

"We see them all the time overhead at night and flying down the intracoastal in the little planes and then those funny ones with stuff on top that make a lot of noise."

"Those are called helicopters," I said. *"Pip's plane is a much bigger one and flies higher and is quieter."*

"Does it fly higher than the birds?" asked Kitty. Her tail moving rhythmically, curling into an S.

"Yes Kitty, a lot higher."

"Oh. That would be exciting," she said.

"Will you tell us again what she said was wrong with Pip. I mean, you know, I rescued him the first time, so I guess I can again. I mean we can," Duke said looking around.

I repeated everything Pip's owner (or mother as we called her) said and answered all their questions.

Tiger's face was all screwed up, like he might be thinking really hard. *"You know,"* he said, *"that reminds me of Georgia when she wanted to call us and couldn't. Remember Duke? Linda said she was acting really strange and restless."*

"Yeah," Duke said. *"Pip's mother said it happened right after his cousin Bo got there. I guess we'll have to go and see."* He looked carefully at Duchess. *"What do you think, Duchess? Should we go?"*

Duchess looked at me, tipped her head, and asked, *"Is it safe in those airplanes? I mean can they fall down?"*

"Yeah?" echoed Tiger with doubt.

"There have been accidents, but not many. A lot more happen in cars, I've flown many times and I'm still here."

"Humph," uttered Tiger.

"Is it scary?" Kitty asked in a little voice.

"It's noisy when it first takes off and then when it lands. But when it's flying it's quiet and you can even sleep if you want." I replied eyeing Duchess. I thought she would be the deciding voice.

Duchess walked over to Duke, rubbed her head against him and said, *"Well, Duke it sure would be a new and different adventure. You wanna go?"*

Duke looked at me. *"You're not afraid are you?"*

I shook my head.

"Then that's settled. If she's not afraid then we can go with her." He looked back at Duchess. *"Might be a lot of fun."*

"Hey hey, what about me? I'm not so sure of this," Tiger said

"You can stay at the vets in a cage till we get back," Duchess said.

"No, no Tiger. You have to come, Pip is more your friend then anyone else's. You found his mother for him," Kitty said.

"That's right Tiger," said Duke. *"Pip would be really upset if you aren't with us."*

"Well maybe," he said.

"I know," Duchess said to me *"You can get him one of those pills, Pip's mother mentioned, or better yet a shot."*

"Come on Duchess," he looked at me. *"You wouldn't do that, would you?"*

I reached over and picked Tiger up cuddled him in my arms rocking him I said, *"No Tiger, no shots and no leaving you at home. You'll be just fine."*

* * *

We arrived at the airport, and Bob met us. He gave all the pets a rub. He asked me if the pets had flown before and I said no, this would be the first time.

He said since they had harnesses on he would snap them to seat belts on take off. I told him I would hold the cats. He offered to hold one of them and Kitty jumped into his arms.

It was a lovely Gulfstream with seats that opened into couches. We settled the dogs on the floor on each side of me. I put my seat belt over Tiger who was sitting on my lap. Bob sat across from us, Kitty on his lap. She was leaning over looking out the window.

"Wow, this is big. Huh?" Tiger said looking around.

Duke looked up at him, *"Not scared any more?"*

"Too late for that," he paused *"Well yeah a little."* He cuddled closer to my chest.

The plane taxied and then raced down the runway, as it took off the dogs leaned tightly into my legs.

When we leveled off Bob said, "They can walk about if it's OK with you."

Tiger had hidden his head under his paws. We heard, *"Are we still alive?" as* he peeked out and looked about.

"Tiger, look out the window I think we live down there some place," Kitty said.

Tiger, Duchess and Duke moved past me, as I unclipped their harnesses. They crowded to the window next to me.

"That's a long way down," said Duchess.

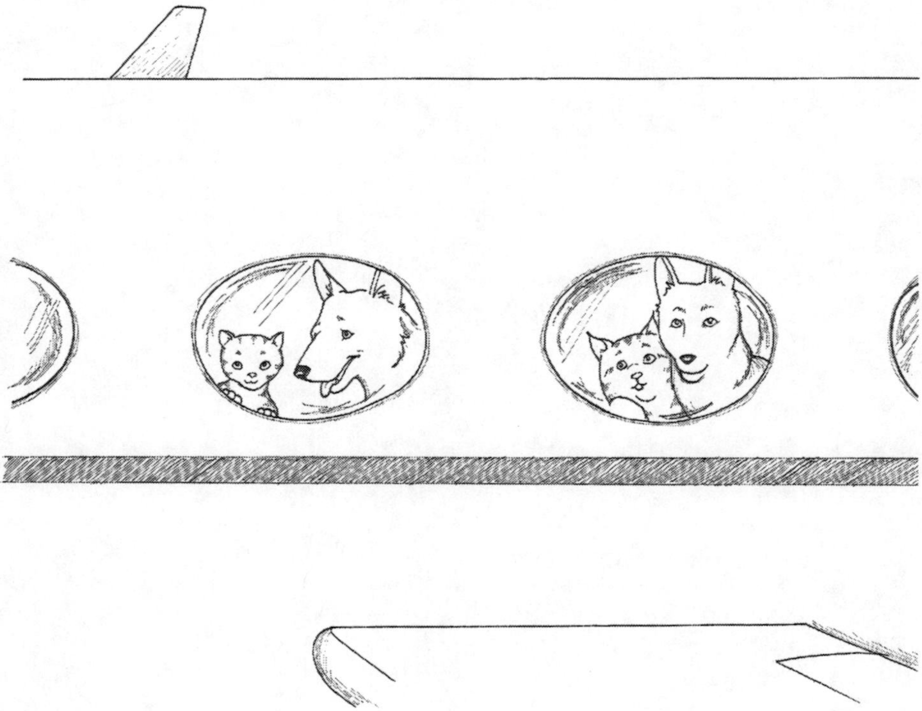

Then they just stared out the window in awe, if such a thing can be said about dogs and cats.

24
Pip and Bo

Pip's home was a rambling two-story weathered gray wooden house trimmed in white surrounded by large covered porches with French doors opening out on them. They called it a cottage. I wonder what a main house would look like. Becca showed us to a large bedroom on the first floor with patio doors opening onto the porch.

"We had dog doors put in here, the library and the kitchen. Do you need a kitty litter box? If so we can fix something up." She looked around and seemed a bit uncomfortable, like the perfect hostess who hadn't thought of everything.

I laughed. "Becca, don't worry. The cats go outside with the dogs."

"Really? I have friends who would love it if theirs could learn to do that. They're afraid they would run away or get lost or stolen or something."

"What? Leave all this. They mustn't be very smart," Tiger said looking around.

" I put some dog and cat beds in here. I mean, it would be OK if you let them on the bed."

"Wow, that sounds good," said Tiger eyeing the bed and me.

"In your dreams," I remarked.

" I was just kidding."

"Don't push it, Tiger" Duchess said. *"I'm sure they could find a special place for you."*

"Can we go out and talk soon?" asked Pip as he paced around.

"Becca, let me unpack and then Pip can take us out and show us the grounds and the beach. Will that be OK?"

"Yes, perfect. I'll see to a few things in the kitchen.."

* * *

We slipped out the bedroom door and Pip showed the cats where they could use the flower garden and we walked across a big grass yard that ran down to the beach. It was lovely, not hot and humid like Florida. The air was fresh and the waves were just lapping on the shore. We .sat on the grass just above the sand when I said, *"OK Pip, tell us what this is all about."*

"Yeah, what's going on Pip?" asked Duke.

Duchess growled. *"Better be good to get us to fly all the way up here."*

"Come on Duchess, give him a chance." Tiger sitting next to Pip nudged him saying, *"Don't worry Pip, that's just the way she is."*

"It's OK Pip." Kitty purred at him, *"Just talk."*

"OK. See that house down there, the second one? That's where Bo and his family are staying for two weeks. Bo raced down to find me right after they got here. He was really upset. Said someone, actually two people, came to his Dad's office, at work to talk with him. They really hated him and wished he was dead."

Duke cocked his head. *"When was this? And how do you know?"*

"Look! Here comes Bo. He'll tell you."

We looked up, a Portuguese Waterdog, with black and white markings like Pip's came running toward us from the beach. He was a large dog. He skidded to a stop, he and Duke and Duchess went through the sniff and greet ceremony. He was then introduced to Tiger, Kitty and me. He wasn't surprised that they could talk with me because Pip had told him all about Florida. Pip told him what he had told us so far and asked him to take over.

"OK Bo, we were asking Pip how you know all this. Were you there?" Duke asked

"Yeah, were you there?" Duchess echoed.

"Yes, sometimes when the children are at school and no one else is home I go over to his office and lie down behind his desk. When people come in, I stay there, unless he invites me out to greet them."

"Does he do that often?" asked Duchess.

"He sort of leaves it up to me. I mean I sort of know when to stay put. If it seems to be sort of important and a formal meeting I stay unless I'm asked."

"What if you want to go to the bathroom?" asked Kitty.

"Oh Kitty, I usually go before I go in or I get my leash and take it to him or one of his men and they give me a quick walk."

"You have walkers?" asked Tiger looking at me.

"He lives in a city and can't be let out alone." I said.

Duchess said, in an attempt to get on with the problem, *"What's with the two people who really hate him?"*

"They came together, a man and a woman one time, and then the man came alone the next time. This all happened last week. A lot of people don't like my Dad but this was different. They acted like they really liked him on the out side but inside they were awful hateful. I'm afraid they will hurt him"

"Where were you when this happened?" I asked.

"I started to stay at the desk but I felt the bad-hate feel and went and sat next to him. I leaned against his legs almost in front of him."

"What did he do?"

"He looked at me kinda funny, then just rubbed my head and went on. I didn't act like I wanted to greet them and he somehow must have known it and let them leave. The same thing happened the second time when the man came alone. I stayed right at my dad's side and almost in front of him," he paused. *"This time I felt like my Dad was glad I was there and he seemed a bit distant from the man."* Again he paused a bit. *"I just don't know how to say this, I've never felt this hate thing. Will you tell my Dad?"*

We were all silent, then I heard Becca call. I looked back at the house and waved. I called, "I'll be there in a minute." She waved a cocktail glass. I waved back. I asked Bo for more details about the man and woman, what they wore, the color of their hair. Then said, *"Bo, let me think how this can be done. Does he walk on the beach with you alone?"*

"You mean with out the family?"

"Yes."

His ears picked up. *"Yes, in the morning early. I'll bring him toward you tomorrow."*

"Since we don't know what time I'll watch for them," Duke said. *"Then I'll come get you."* He looked at me. *"OK?"*

* * *

Before bedtime I went for a stroll with the dogs down the road in front of the house. I had put on a light jacket that Becca loaned me. She said I would find it much cooler at night then at home. The road was a small two-lane road with tall trees along each side. The homes lay at the end of drives that curved through the trees. It was quiet and very private. Duchess nudged me and asked, *"Do you know what you're going to do yet?"*

"I'm not sure. But like Bo, I think he needs to know. People can hurt you in many ways, especially if you aren't expecting it. "

"Tell him the truth," she said.

"You mean about us?" Tiger, walking along side Duchess, asked.

"I think so. What's the harm, all he can think is that she's nuts," replied Duchess.

I smiled, and thought. Well why not. If he told anyone they would think *he* was nuts.

Duke got up early the next morning. I wasn't far behind. I got a mug of coffee and headed to the beach. Bo and his Dad were walking along the surf line. Pip barked and ran to them. He and Bo nudged his dad toward us. I stood, and we walked toward them.

After introductions he waved back the men who were following and yelled, "It's OK." Then he said, "Becca told us about you rescuing Pip. You can bet we made sure Bo was chipped." We walked and talked a bit about animals and the lovely weather.

"Tell him, tell him." said Bo and walked over by me and nudged me.

"Hey Bo what's going on?" his Dad asked.

"Don't blow it, Bo," said Duchess. *"She'll get to it."*

I squatted down by Bo and said, "Do you mind if we stop a few minutes? I'd like to talk to you about Bo."

"About Bo?" His eyes grew large. "What about Bo?" We sat down on the sand leaning against a drift wood log.

"I don't know how to put this without you thinking I'm completely crazy. Bo told us he's afraid for your life."

"Well that was sure putting it all out there," said Tiger.

Bo nudged his Dad and nodded his head. He looked at Bo and then at me. "He told you this. How did he do that?"

I then told him about my telepathic ability with the animals. He listened and seemed ready to stand up when Bo pushed him again. *"Tell him about those two people and what I told you."*

So I quickly explained what Bo had told us. "He also said the woman wore a red dress and had black short hair. The man had longish blond hair."

He sat and stared at me, at the animals and at Bo. "I think *I'm* going nuts." He shook his head. "It's this damn job," he paused, "but you're right. It did happen just as you said or rather as Bo said, I mean as you said…."

I laughed and explained, "I thought I was losing it when Duchess first spoke to me at the shelter." I told him how I tested her. "Try it. Just whisper to me what you want them to do. I'll mentally tell them to do it."

The poor man just looked at me and shook his head again. He then whispered in my ear what he wanted them to do. He looked at Duke who got up and walked all around us and sat. Duchess then did the same. Tiger got up, came and sat in his lap, and Kitty sat in mine. Bo and Pip sat still. "Hum" he said. He whispered some more. They all got up and stood in front of him in a nose to tail line, with Bo and Pip at the rear. "I'll be damned." He shook his head in amazement. "What else can you tell them to do?"

Since Duke and Duchess had harnesses on I said, *"OK Kitty up on Duke and Tiger on Duchess. Run down to the surf and back."* They did. *"Now Kitty you get on Bo and Tiger on Pip and do the same."* I heard, *"No claws,"* said by Bo and Pip at the same time as Kitty and Tiger climbed gently up. They walked more slowly to the surf. I laughed and said out loud," Bo and Pip don't quite trust the claws."

I turned to him, "We all talk to our pets and many people will tell you that they understand and even talk back in their own way. This is just a step or so further. Also German Shepherds, your dog and some others are very smart. So are cats. It seems the more you talk with them the better the communication gets."

"OK I'm convinced, sort of." His furrowed brow and squinted eyes reflected his thoughts.

"Tell him he must keep it a secret," Duchess said very emphatically nodding her head, the others nodded as well

"What?" he said.

I repeated what Duchess said and explained why we all felt that way.

He quietly nodded. "I don't think you have to worry about me telling anyone, humph. First no one would believe me, and second I'd never get reelected. About those two people, now that I think back about them, I clearly remember Bo acting funny and the second time I was glad he stayed close to me. Funny, my wife knows them and told me she doesn't like either one of them. She said, quite clearly, she didn't trust them. Now I can tell her Bo didn't like them either."

"No, No. You can't tell her either." They said in unison, standing up looking alarmed. Bo continued, *"She might tell the girls."*

"What's wrong?" he said looking around.

I explained what they and even Bo had said. "It's hard to keep a secret like that."

"Let me ask, can they tell what everyone is thinking?

Duke and Duchess looked at each other. *"Well?"* I asked.

"No, not thinking but if some one has or directs a strong feeling right at us or near us, we can pick it up. I think it's because, well… I don't know why." Duchess replied

I told him what they said. "They don't spend a lot of time with people."

"Except on the golf course," quipped Tiger.

I repeated what Tiger said. *"Tell him about golf,"* Tiger said.

So I did. He laughed, "I don't think we could do that here," he paused. "Well I could ask maybe before it opens and no one is around you could hit a few also."

"Can you help him talk to me and listen to me?" asked Bo.

"Yeah, use sign language. I saw that on television," said Tiger. *"Tell him."*

I told him what they said and added we would only be there another couple of days. "We could start practicing later today. Bring the girls to meet the others this afternoon. They can play catch with the dogs while you Bo and I sit on the grass and practice. That OK?"

"Sounds great. My wife said we were invited for drinks and dinner later this evening at Becca's and Bobs. But I'm sure glad Bo had us meet this morning." He stood, "No, I don't think you're nuts, and yes I keep secrets very well. Come on Bo." They jogged off down the beach.

"That went well," said Duchess.

"Sure did," echoed Kitty and Duke.

"Yep, we're really good," said Tiger and Pip nodding.

Then Tiger looked at me. "I *knew you could do it."*

Looking up I said, *"Thank you, Lord."*

I heard, *"Amen."*

25
Thank You! (You're Welcome)

After lunch Becca went to the village to shop and we went to the beach to wait for Bo and his daddy. *"Here they come,"* yelled Pip. *"Oh, Oh, the girls and Bo's mom is with them."*

Bo came running up to us and between pants told us, *"Don't worry. They're going for a long walk and run on the beach. Dad told them to bring balls to throw for the dogs."*

After we were all introduced, Bo's dad said, "Why don't you guys take the Shepherds and Pip with you for your run. I'll stay here. We can talk about golf and the condition of the world."

I laughed, "The first would be easy, I'm not so sure about the second."

"Neither is my dad," quipped Bo.

I looked down at him in surprise, "Bo?" I said out loud. He hung his head.

"What?" His dad queried

"Nothing," I smiled. Duke, Duchess, and Pip were dancing around ready for a run.

* * *

"How do you want to start?" he asked.

"Why don't you tell us what verbal commands you use with Bo and does he obey?"

"What? Of course I obey. They sent me to school."

"What's with this school business?" Tiger asked, looking at Bo's dad.

"Tiger wants to know what's with the school business."

He looked at Tiger, sighed and answered, "When we were going to get Bo we had him first go to dog training school. Then the trainer showed us what we had to do with him."

Tiger nodded.

I suggested he put Bo through each command. They went through all of them. Bo followed each one promptly. Things like, sit, lie down, stop, come, even go and a few others. "See he's a good dog and really very smart."

"Do you ever not obey?" Tiger shrugged. *"You know if you don't want to do something."*

"Well, sort of. I may act like I didn't hear them the first time or I may lay when they want me to sit. That sort of dumb stuff." He paused, looked at me *"You won't tell him will you?"*

"No. We'll talk about that later."

"Tiger watches a lot of TV. He thought you could develop some sort of sign language between you. Stuff you could both use to get the others' attention."

"Like simple yes and no questions and he could nod or shake his head." Tiger stood up, *"Tell him."*

I repeated what Tiger said.

"OK, I got it. Lets start. Are you hungry, Bo?"

Bo shook his head no. "Are you thirsty?" Bo shook no.

"Why not?"

"Oops," Kitty said, *"first mistake."* She jumped up in my lap.

Tiger walked over, stood up and tapped him on his knee, shaking his head, no. *"He does know what yes and no means, doesn't he?"*

I laughed, explained what they said and what he had done wrong. He looked at me and then at Tiger. "Tell you what, please just give me an ongoing translation of everything they say."

"Everything? Even the smart ass remarks?"

"Especially those." He eyed Tiger. "It makes it easier, cut right to the chase without embellishments."

"Emb… what?" asked Tiger.

I explained what embellishments meant.

"Oh, guess I had better watch my mouth." I repeated this. Kitty just started washing her face.

"No, tell him to just be himself."

"OK, Let start with two things. First tell him to do that 'tsk-tsk' sound with his tongue like you do, to get Bo's attention. Show him."

I demonstrated the sound and how to make it. I told him to practice various volumes at different distance. He looked skeptical and Tiger started to speak… I interrupted and suggested he just do it.

He started to speak, then shrugged and did it.

"Oh boy, is he going to question everything?"

"Probably." I replied.

"He does that with everybody, seems to come with the job," Bo said.

He raised an eyebrow, shrugged, and 'tsk-tsked' with Bo. I'm sure he felt foolish but then he got the hang of it. Bo was very responsive with the head nodding and shaking. They worked on both the volume and distance.

"OK," said Tiger, tapping him on the knee and taking command again. *"Lets move on. Teach him how to use hand commands, which one and why. Then switch to only using a finger."*

I explained all this and demonstrated. He paused, looked at me, then Tiger. Smiled, "Got it."

How clever Tiger is, I thought. Maybe the TV watching isn't too bad after all.

Tiger watched very carefully nodding with each success. *"OK, that's enough, you're doing good. Lets move on,"* he said, tapping him on the knee again.

I explained that Tiger wanted to move on. He replied "I think we should practice more…"

Tiger interrupted him and this time I quickly translated. *"You can do that on your own time, we need to move on."*

I couldn't help myself. I was grinning. Bo's dad looked at me, "You're really enjoying this, aren't you?"

"You asked for a literal translation, I warned you," then I really grinned and he laughed out loud.

"That goes for you too," Quipped Tiger. *"We need to work on the protection stuff and the warnings."*

I told him that Tiger wanted to work on protection stuff. I said to them both I didn't think that was appropriate. That Bo's dad had men there to protect him and Bo was not a guard dog.

"What about the moves, Bo did in his office with those two people?"

"Tell him please," said Bo.

"OK, Bo." I told him again about the two times Bo had gone to stand

half in front of him, when the man and woman were in his office. I said that Tiger and Bo both thought some sort of signal was needed to warn him of really hostile people.

Tiger was in his lecture posture, sitting tall with a paw on his hip. I asked if Bo greeted everyone who came into his office.

"Of course not. My office is a place of business. It's OK if he's there as long as he stays behind my chair, unless I invite him out." He paused. "Sometimes he comes out because he knows the people and they like to see him for a minute."

I explained again what Bo had done and asked if he remembered.

"Well sort of. Can Bo tell the difference between someone who just doesn't like me much and some one who really hates me?"

"Bo, do you understand the difference?" I asked

"I think so. I won't forget that real hate thing." He paused. *"Tell him I won't make a pest of myself."*

I repeated what Bo had said.

"Come, Bo." he leaned down and hugged him good. "You will never be a pest to me, Bo."

"Oh, Oh, here come the others." Tiger nodded toward the shore.

"We'll be leaving after lunch tomorrow. Would you like one more quick practice session early in the morning?"

"Yes. Becca said we are all invited to dinner tonight. You guys are too." He nodded to the animals.

Tiger and Kitty nobly nodded as if they knew this all along.

I rolled my eyes. I was doing that a lot lately.

* * *

After dinner we were all sitting in the living room when Bo's mom asked her husband "What were you two doing this afternoon?"

He looked at me. I said, "We were teaching hand signals, for Bo to follow. He spends time in your husband's office. It seems like a good idea for them to learn to communicate, silently when people are there."

"Oh wow. Dad can we learn too?" the youngest daughter asked and off we all went on a conversation about talking with your dogs.

Duchess was lying at my feet. *"You know this might work out just fine."*

"I think so too."

"But I wouldn't trade being able to actually talk with you."

I reached down and rubbed her ears, *"Me either"* said Duke.

* * *

The next morning after Bo, Tiger, Kitty and I did more work on hand signals. Soon the dogs, kids and his wife came back from a long run on the beach. Bob and Becca joined us and tried hand signals with Pip and yes and no questions. "I don't understand how he knows so much."

Bob looked at his watch. "Time to catch the plane." Everyone said goodbye.

Then Bo's dad picked up Tiger hugged him and told him he was the smartest cat in the world, next to Kitty. He thanked them both and said he hoped to meet them again.

As Bo and his family walked away, we heard, *"Thank you, Mr. President"* Bo's dad spun around and looked back. Tiger was standing tall with his paw held to his forehead.

The President looked at him, snapped a smart salute back and said, "You're welcome, Tiger."

The Author

Doctor Kohnke, while a Professor at New York University, spent a good part of each year at her home on the Intracoastal Waterway in Palm Valley, Florida. The love of the waterway lured her into early retirement. She began a new lifestyle, becoming active in her community when she wasn't fishing, working in her large yard, and tending her German Shepherds, until she was elected as a County Commissioner,

She is an advocate of the environment and has worked to preserve the wetlands, birds, animals and trees. She fears the rapid growth in Florida will leave little of its charm for future generations. After her term in office she again retired to her fishing increased her golfing and began sharing her love of Florida and all its inhabitants, even the two-legged ones, in these fictional stories.

CPSIA information can be obtained at www.ICGtesting.com
Printed in the USA
LVOW130909200812

295082LV00001B/2/P

9 780977 453658